DATE DUE FOR RETURN

D1349687

This book may be recalled before the above date

HIGH
PERFORMANCE
MANUFACTURING

Wiley Operations Management Series for Professionals

Harnessing Value in the Supply Chain: Strategic Sourcing in Action, by Emiko Banfield

The Valuation of Technology: Business and Financial Issues in R&D, by F. Peter Boer

Logistics and the Extended Enterprise: Best Practices for the Global Company, by Sandor Boyson, Thomas M. Corsi, Martin E. Dresner, and Lisa H. Harrington

Total Project Control: A Manager's Guide to Integrated Project Planning, Measuring, and Tracking, by Stephen A. Devaux

Internet Solutions for Project Managers, by Amit K. Maitra

Project Management Success Stories: Lessons of Project Leaders, by Alexander Laufer and Edward J. Hoffman

High Performance Manufacturing: Global Perspectives, by Roger G. Schroeder and Barbara Flynn

HIGH
PERFORMANCE
MANUFACTURING

Global Perspectives

ROGER G. SCHROEDER
AND BARBARA B. FLYNN
Editors

John Wiley & Sons, Inc.

New York • Chichester • Weinheim • Brisbane • Singapore • Toronto

Contents

Contributing Authors

Kimberly Bates   University of Toronto, Toronto, Canada

Kate Blackmon   University of Bath, Bath, United Kingdom

Alberto De Toni   University of Udine, Udine, Italy

Roberto Filippini   University of Padova, Padova, Italy

Cipriano Forza   University of Padova, Padova, Italy

Barbara B. Flynn   Wake Forest University, Winston-Salem, North Carolina

E. James Flynn   Wake Forest University, Winston-Salem, North Carolina

Frank H. Maier   University of Mannheim, Mannheim, Germany

Yoshiki Matsui   Yokohama National University, Yokohama, Japan

Peter M. Milling   University of Mannheim, Mannheim, Germany

Aneil Mishra   Wake Forest University, Winston-Salem, North Carolina

Kelly Mollica   Wake Forest University, Winston-Salem, North Carolina

Michiya Morita   Gakushuin University, Tokyo, Japan

Sadao Sakakibara   Kanagwa University, Yokohama, Japan

Osam Sato   Tokyo Keizai University, Tokyo, Japan

Roger G. Schroeder   University of Minnesota, Minneapolis, Minnesota

Kathrin Tuerk   BASF AG Lundwigshafen, Germany

Andrea Vinelli   University of Padova, Padova, Italy

Chris Voss   London Business School, London, United Kingdom

Sven Weissmann   University of Mannheim, Mannheim, Germany

PREFACE

This book is the result of a long and arduous journey. It began in 1989 with a research project aimed at collecting data to understand the emergence of Japanese factories in the United States. At that time, we believed that the best U.S. plants could match the Japanese transplants in practices and performance, but that the average U.S. plants would not compare. To test this idea, an initial group of researchers from the University of Minnesota and Iowa State University—Kim Bates, Barbara Flynn, Jim Flynn, Sadao Sakakibara, and Roger Schroeder—conducted round one of data collection.

Soon, we obtained interest from university professors in other countries who also wanted to participate in our study by collecting comparative data in their respective countries. This led to round two of the data collection. For round two, we substantially revised the questionnaires and procedures based on what we had learned in the first round. Measurement was vastly improved so that we could accurately represent the demographics, practices, and performance of each plant. In round two, the questionnaires were written in English and translated into Italian, into Japanese, and German so the data could be collected in the native language of the countries involved. The group of professors in each country took responsibility for the data collection by contacting the plants in the sample and then visiting them to collect data. They spent many hours ensuring that the data were accurate and that it represented what was actually happening in the plants studied. As a result, we believe that the data represents the most thorough and valid set of data on global manufacturing today.

In addition to the data collection efforts, we also enjoyed meeting with the research group in various locations and countries around the world. Over the course of six years we have often met once or twice a year, sometimes in resort locations and sometimes in the home countries of the authors, to work on book chapters and to debate issues affecting global high performance manufacturing (HPM). These meetings started with a research agenda, which then evolved into a book plan. We felt that what we were learning would be of interest to managers as well as to academics and should be written in a format that would be easy to understand by the practicing management community.

ix

After reading Chapter 1, you will realize that this is not another "how to" book. We hope that it represents an awakening and a turning away from "the flavor of the month" approach that is so prevalent today. We are encouraging plants to set out their own path to global high performance manufacturing, depending on their particular country, industry, strategy, and situation. Each plant is different and should adopt practices in the order and to the extent that they fit the plant's personality and context.

We also stress the idea that linking practices together is essential to global high performance manufacturing. This idea of integration is critical and most difficult to achieve in practice. Often new practices are tacked onto existing practices and not carefully linked into the fabric and the infrastructure of the plant. As a result, when the plant manager or practice champion leaves the plant, the practice will die.

This book provides a glimpse into how just-in-time (JIT) practices, total quality management (TQM), new technology, information systems, human resource practices, and manufacturing strategy should be introduced into a plant environment to achieve high performance manufacturing. It takes a comprehensive view by looking at all practices that can be implemented and by showing how to select and integrate them into the plant's particular situation. As a result, a plant can determine the set of practices that actually fits its environment best and not attempt to implement practices that are not needed to compete.

This is one of the first books to provide comparisons of manufacturing in the United States, Japan, Germany, Italy, and the United Kingdom. We provide a chapter on each of these countries to represent the unique situations as seen by local experts using our common framework. As a result, current manufacturing issues in each of these countries are addressed along with insights that affect global manufacturing anywhere in the world.

This book aims to set a new standard for the way manufacturing practices are viewed in today's world. To the extent it helps managers more clearly address the situations facing them, it will be a success. We wish the readers the best of luck in achieving global high performance manufacturing in their company.

Roger G. Schroeder
Minneapolis, Minnesota

Barbara B. Flynn
Winston-Salem, North Carolina

ACKNOWLEDGMENTS

$\mathbf{W}$e would like to acknowledge the hard work of various students who helped with data collection, data cleaning, and data analysis. At the University of Minnesota these students were David Hollingworth, Sohel Ahmad, Mikko Junttila, Kristy Cua, Sarvanan Devaraj, and Manus Rungtusanatham. At Iowa State University, they included Donna Gilligan, Mike Walters, and Connie Blair.

We would also like to thank the managers and employees from the 164 plants that participated in our study. Without their willingness to share their management practices, plant performance and philosophies this book would not be possible.

Our Editor at Wiley, Jeanne Glasser, was extremely helpful in planning the manuscript and guiding us through the editorial process. Publications Development Company of Texas was very helpful with the production process.

Finally, we would like to thank the National Science Foundation which funded some of the research underlying this book. An early grant was also provided by the Japan-U.S. Friendship Commission that greatly assisted us in data collection. We would like to thank the State of Baden Württemburg, Germany and the German research foundation, DFG, for their financial support.

We are grateful to our families for the many long hours they sacrificed while we were working on various phases of the book. Without their support and encouragement this book would not have been possible.

PART I

SETTING THE STAGE

CHAPTER 1

HIGH PERFORMANCE MANUFACTURING: JUST ANOTHER FAD?

ROGER G. SCHROEDER and BARBARA B. FLYNN

$\mathbf{T}$ony Salvatori gazed out the window of his plant office in Milan, Italy, as he contemplated the events of the past few days. An American consultant had just left his office after having recommended that his plant adopt a Six Sigma quality improvement approach. With great passion the consultant told of companies in the United States that had achieved great success with Six Sigma, including such giants as Motorola, General Electric, and Seagate. While Salvatori saw the merits of this approach, he wondered if it would fit the personality of his plant. Would his engineers and managers accept the statistical methods that are required to implement the seven tools, process control charts, and design of experiments? Would these techniques actually help to improve processes in his plant? And was his plant ready for Six Sigma? Of course, the consultant had not addressed these questions but had argued that everyone should be using Six Sigma in a modern plant such as Tony's. What worked in the United States would work in Italy, too.

Today, manufacturing managers are being hit from all sides with the latest fads—from total quality management (TQM) to just-in-time (JIT) to business process reengineering (BPR) and now Six Sigma. But this book is different; we don't seek to create yet another fad. We advance the idea that the management practices leading to global high performance manufacturing differ by country, industry, and size of company, to name just a few contingencies. We take this contingency approach to adoption of management practices rather than a universal, one-size-fits-all approach. Thus, the first distinctive feature of high performance manufacturing is that it uses a *contingency approach*.

3

Our contention is that managers should carefully assess their unique environment and strategy before adopting the latest management practice. What works in one country and situation may not work elsewhere because of different national cultures, economic conditions, or competitive environments. Thus, to achieve high performance manufacturing (HPM), best practices must be selected and adopted relative to the situation the company faces. However, managers must not use the contingency argument as an excuse to avoid change. They should carefully study the new approaches; try them on a pilot basis, when appropriate; then determine whether the approaches should be fully deployed, or look for adaptations that might be appropriate. Rejecting new approaches out of hand is a risky business, just as is adopting every new approach that comes along.

It is well known in the academic literature that the strategy adopted and the associated practices should depend on the environment of the organization. This fact seems to have escaped many consultants and practitioners who argue that the latest management practices should be adopted by everyone. Aside from the fact that what is touted as new is not always that new after all, the advocates of universal best practices have been overselling their approaches. This had led to a series of failed initiatives and to cynicism that the latest new idea is just another fad. When viewed in a contingency perspective, it may not be that the fad failed but that management failed to apply the latest ideas in the appropriate situations. This book uses a contingency approach to provide practical advice for managers who face complex situations and who find they must select and adapt the approaches they use to their particular situation.

Returning to Tony Salvatori once again, it wasn't just Six Sigma that was bothering him. He had faithfully implemented TQM, beginning 10 years ago, and was continuing to build on his quality improvement efforts. The workforce had undergone intensive training aimed at employee involvement and teamwork. About five years ago the plant began implementing JIT by reducing setup times, stabilizing the master schedule, forming manufacturing cells, and cross training workers, to name just a few of the changes. Recently, several managers from the plant attended a seminar on enterprise resource planning (ERP). So what should Tony's next move be—enterprise resource planning, Six Sigma, or neither? Tony realized that what was also bothering him was that he had not fully linked the past practices together into a coherent whole. If he adopted one of these new ideas, how could it be linked to the practices that he already had in place? Should he concentrate on linkages when starting something new?

A second distinctive feature of this book is the emphasis we place on *link-ages* among practices. We argue that linking one practice to another is what leads to HPM. It is not that success follows just from the number of practices that are implemented or from the latest practices, but from how they are re-lated to each other and how they cumulatively build on one another. While this makes eminent common sense, managers that we have interviewed tell us that establishing and maintaining linkages across all of their new and old initiatives is one of the most challenging problems they face.

Linkage of practices provides the basis for tying new initiatives into what the plant is already doing. And integration among practices is an ongoing need that must be continually renewed. Once a linkage is achieved, it be-gins to deteriorate; and maintaining it requires constant attention.

But what is *linkage?* Just another slogan or is it something that is truly fundamental to management? Let's return to Salvatori's plant. He has a fully functioning quality management system, and suppose, for the mo-ment, he wants to add Six Sigma. What would be the best way to do it? Should Tony set up a separate group of black-belt analysts who would im-prove processes anywhere in the plant, as the consultant recommended? Should the analysts be trained in all of the standard techniques and who should do the training? In other words how should the new program be im-plemented to link into the practices that already exist?

In this case Tony's plant does not have a separate quality department be-cause this is not a common practice in his company. Rather, quality con-trol and improvement are delegated to the line workers and engineers. Setting up a separate department or group of "black belts" would run counter to the existing plant culture and would shift attention to the staff rather than to the line responsibility for quality that already exists. Also, the plant is quite advanced in its use of statistical process control and the seven tools of quality for process improvement, but it has rarely used design of ex-periments to improve quality. Therefore, maybe the Six Sigma effort in Salvatori's plant should be directed at teaching design of experiments to engineers and lead workers, who would then use it to further improve processes. Salvatori could also ratchet up the rate of process improvement by using existing tools and personnel in place. This is an example of link-age of a new practice to the present situation and context. A program such as Six Sigma should not be implemented in isolation but should be linked to practices already in place. Furthermore, the application of any new pro-gram depends on the context of the plant in terms of its size, industry, and country.

THE HIGH PERFORMANCE MANUFACTURING DATABASE

A third distinctive feature of this book is that it is based on a well-defined and structured database including 164 plants around the world. We will briefly describe the data that we have collected in order to set the stage for the remainder of the book, where the data is routinely used and compared across plants and countries.

The data were collected by using a standard set of questionnaires together with site visits to many of the plants. The questionnaire packet addressed six areas of plant management practices: (1) manufacturing strategy, (2) TQM, (3) JIT, (4) human resources (HR), (5) information systems, and (6) technology management. By covering these areas, we conducted a complete audit of the plant's management practices to determine the extent of implementation of the various practices. We also measured the performance of the plant in both absolute and relative terms compared to its competitors and its global industry, and we evaluated the context of the plant in terms of its environment, size, industry, and so forth. The methods used to measure the many practices that we studied in each plant are described in Appendix 1 at the end of this chapter.

The selection of the plants to participate in the study was based on several criteria. First, about half of the plants were randomly selected from lists of "high performance reputation" plants that had been touted as leaders in the literature or by industry experts. This was done to ensure that the sample contained a good representation of some of the best plants in the world. The other half of the plants were selected at random from lists of the general population of plants. This provided a comparison group consisting of the more traditional and ordinary plants. The selection also included plants from three industries in each country: (1) electronics, (2) machinery, and (3) automobile component suppliers.

After a plant was selected to participate in the study, a member of the research team personally contacted the plant manager to request participation in the study. As a result of this personal contact we obtained a two-thirds response rate, which is quite high for a study of this type. The result of this effort is shown in Table 1.1, with the number of plants surveyed by type in each country. See Appendix 2 for a list of some of the companies included in this study.

As a result of the aforementioned efforts we have a unique and valuable database of practices adopted by manufacturing plants around the world

Table 1.1
Numbers of Plants in the HPM Study

	United States	Japan	Italy	Germany	United Kingdom	Total
High performance reputation	14	32	18	15	2	81
Traditional (ordinary)	16	14	16	18	19	83
Total	30	46	34	33	21	164

and the associated plant performance and context of the plants. The implementation of these practices will be examined in this book along with the linkages among practices and the context of the plants that we believe explains why a plant has adopted a particular practice in a certain way. This is the first book to provide this perspective and information for managers.

THE PROPOSED HPM MODEL

The proposed model that underlies this book is described in more detail. First, we define *high performance manufacturing* as consisting of the following six practice areas:

1. Manufacturing strategy
2. Total quality management
3. Just-in-time
4. Human resources
5. Information systems
6. Technology management

This particular model is quite broad based when compared to models that have been proposed in the past. For example, Hayes and Wheelwright (1984) suggested one of the first models of high performance (or world class) manufacturing. They identified the practice areas as:

Build the skills and capabilities of your workforce.

Build technical competence through management.

Compete through quality.

Develop real worker participation.

Rebuild manufacturing engineering.
Develop breakthroughs and continuous improvement.

Their model was based on close observations of what Japanese, German, and the best U.S. manufacturers were doing at the time. Our proposed model includes all of the Hayes and Wheelwright practices and plus JIT and information systems, which were not included in the Hayes-Wheelwright model. Furthermore, we have broadened their areas of technology management, quality management, and human resources to include several more variables.

A second example of world class manufacturing is the model proposed by Schonberger (1986) and subsequently expanded by him (Schonberger, 1990, 1996). In his 1986 book Schonberger argued that world class manufacturing consisted of JIT, TQM, employee involvement (EI), and TPM (total productive maintenance). Once again, our model broadens Schonberger's notion of EI and technology management to include additional practices and adds manufacturing strategy and information systems to his list of practices.

We make these comparisons not to criticize the contribution of these authors, but merely to point out the broad nature of our approach that is inclusive of many of the practices included in previous studies concerning HPM.

The linkage idea that we stress in this book can be seen in Figure 1.1, in which we indicate by overlapping circles that JIT, HR, TQM, information systems and technology management practices should be linked together. All of these practices in turn should be guided by manufacturing strategy to link the plant to its external environment. The external environment consists of political, economic, social, and national forces. These environmental forces are in constant change and require adaptation and selection of the practices used by the plant to meet the changing situation.

But what do we mean by *internal linkage between practices?* As explained earlier, it means that practices are linked together over time and that new implementations consider what has already gone before. As a result, the practices tend to reinforce each other and provide synergy. A plant that has a well-integrated set of practices guided by an overall manufacturing strategy is a joy to see in practice.

The linkage to the external environment can be thought of as the contingency approach that we have been discussing. As we shall see in Chapter 2,

Figure 1.1
HPM Model

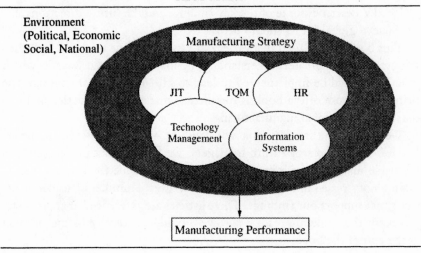

Environment
(Political, Economic
Social, National)

Manufacturing Strategy

JIT   TQM   HR

Technology
Management   Information
Systems

Manufacturing Performance

the paths followed to HPM in different countries have varied greatly, thereby supporting our argument of contingency. The environment is a powerful force and cannot be ignored in selection and implementation of practices.

ORGANIZATION OF THIS BOOK

The book is organized into four parts. Part I begins with this introduction to our approach. In Chapter 2 we describe the contingency approach in more detail by examining the unique paths that plants in different countries have taken to achieve HPM. In Chapter 3 we discuss the linkages that must exist between practices to achieve high performance.

Part II explains the six sets of practices that constitute the content of HPM. Each chapter is devoted to one of these six sets of practices. Comparisons are made across countries, and the contingencies and linkages for each practice area are discussed.

Part III describes the practices adopted by specific countries. For example, Chapter 13 explains how small- and medium-size companies achieve high performance in Italy because there are very few large companies in Italy. As we shall see some unique adaptations of practices occur because of the size of the firms in Italy. Chapter 14 describes how the economic and political environment of Germany has affected the practices used in that country. Chapter 11 strongly supports the Japanese linkage of practices to each other.

Part IV describes some of the key findings and the future directions that we believe global HPM should take.

This book can be used in a variety of ways. We hope the book will contribute toward a greater understanding by managers of the ways in which practices should be implemented to lead to HPM in a global context. The primary purpose of the book is to enhance management practice and understanding of global manufacturing.

We also hope that the book will find its way into classrooms and libraries. As teachers of future managers, we believe that professors need teachable material that is grounded in research. While the research methods are not stressed in this book, we have written a number of academic articles that support our findings. These articles are referenced at appropriate places for the academic readers and we provide a complete listing of them at the end of the book.

REFERENCES

Hayes, Robert H., and Steven C. Wheelwright. *Restoring Our Competitive Edge*. New York: Wiley, 1984.

Schonberger, Richard J. *World Class Manufacturing*. New York: Free Press, 1986.

_____ . *Building a Chain of Customers*. New York: Free Press, 1990.

_____ . *World Class Manufacturing: The Next Decade*. Falls Church, VA: APICS, 1996.

APPENDIX 1

LIST OF SCALES AND OBJECTIVE MEASURES OF HPM PRACTICES

To accomplish the measurement of the practices, we first defined each of the six practice areas by a series of multi-item scales and objective questions. Each multi-item scale consisted of several perceptual questions (items) in which the scores were added to arrive at a total for the scale. For example, we measured JIT by the extent of implementation of Kanban, repetitive master scheduling, daily schedule adherence, and setup time reduction, to name a few of the scales. Each of these scales in turn consisted of several questions to provide measurement accuracy, answered on a scale of 1 to 7.

We also used objective measures in this study. For example, we asked, "What is the percentage of common parts among all products?" Another example was "the number of suppliers to the plant." The following lists describe the scales (not the items or the questions) and objective measures used in the study.

Once the questions and scales were defined, they were allocated to various questionnaires that were given to the plant manager, to several other managers, and to workers. In all, each participating plant was asked to complete a battery of 23 questionnaires, which were given to various individuals (managers, staff, and direct labor). In addition, plant visits were made to many of the plants by one or more members of the research team to observe the practices in use and to discuss them with plant management.

Manufacturing Strategy

Anticipation of new technologies

Communication of manufacturing strategy

Formal strategic planning

Functional integration

Note: Each of the scales and measures includes several questions that are available from the editors.

Long-range orientation
Manufacturing-business strategy linkage
Manufacturing strategy strength
Product competitive performance comparison
Proprietary equipment

Total Quality Management

Continuous improvement
Customer involvement
Customer satisfaction
Feedback
Maintenance
Process control
Quality in new products
Rewards for quality
Supplier quality management
Top management leadership for quality
TQM link with customers
Quality approach
Supplier quality level

Just-in-Time

Accounting adaptation to JIT practices
Comakership
Daily schedule adherence
Equipment layout
JIT delivery by suppliers
JIT link with customers
Kanban
Material requirements planning (MRP) adaptation to JIT

Pull system support

Repetitive nature of master schedule

Setup time reduction

Small lot size

Fixed production schedule

Human Resources

Centralization of authority

Commitment

Coordination of decision making

Pride in work

Compensation for breadth of skill

Documentation of shop floor procedures

Employee suggestions

Incentives for group performance

Management breadth of experience

Manufacturing/human resource fit

Multifunctional employees

Recruiting and selection

Rewards/manufacturing coordination

Shop floor contact

Small-group problem solving

Stable employment intention

Supervisory interaction facilitation

Task-related training for employees

Worker's breadth of job

Compensation ratio

Compensation/rewards/incentives

Egalitarian index

Employee turnover

Training

Information Management

Accounting

Benefits of information systems

Coordination with corporation

Dynamic performance measures

External information: supplier quality control

Internal quality information

Management vision of information technology (I.T.)

Manufacturing plans

Performance feedback

Stability/predictability of short-term production

Applications of I.T. and architecture

Future expenses

Information for JIT suppliers and customers

New accounting systems

Seven tools

Shop floor planning and control technique

Software architecture

Telecommunications systems

Technology Management

Effective process implementation

Interfunctional design efforts

New product introduction process

Product design simplicity

Working with technology suppliers

Automation level

New product introduction cooperation

Willingness to introduce new products

New product development

Plant Performance

Competitive performance on cost, quality, delivery, and flexibility

Accounting data on costs, scrap, rework, on-time delivery, and cycle time

Cost of poor quality

Plant Environment

Complexity of environment

Plant description

Plant focus

Products, parts, and processes

Industry and country

APPENDIX 2

This appendix contains the names of some of the companies/plants that participated in the study. One plant from each of these companies was selected for data collection. The names are listed to illustrate the types of companies that are included in the data set.

United States

Aisin USA Manufacturing, Inc.
Calcom
Caterpiller
Douglas Autotech
Dowling Engine Cooling (Valeco)
Duff Norton
Eaton Corporation
Exide Electronics Corporation
Extrude Hone
Gates Rubber Company
Henry Filters
Honeywell
Hutchinson Technologies, Inc.
Indresco
Intergraph Corporation
John Deere
Komag, Inc.
Prince Corporation
Roper Whitney of Rockford
Signet
Stanadyne
Stone Construction Equipment, Inc.
Telex Communications, Inc.
Tennant
Texas Instruments
Tower Automotive
Unisys
United Technologies Automotive
Verbatim Corporation
ZF Industries, Inc.

Japan

Aisin AW Industries Company Ltd.
Akebono Brake Industry Company Ltd.
Amada Company Ltd.
Anest Iwata Corporation
Casio Computer Company Ltd.
Chuomusen Company Ltd.
Darkin Industries Ltd.
Fujitsu Limited
Hitachi Construction Machinery Company Ltd.
Hitachi Ltd.
Honda Motors Company Ltd.
Ishikawajima-Harima Heavy Industries Company Ltd.
Isuzu Motors Ltd.
Keeper Company Ltd.
Komori Corporation
Kubota Corporation

Kuroda Precision Industries Ltd.
Makino Milling Machine
  Company Ltd.
Mazda Motor Corporation
Mitsubishi Electric Corporation
Mitsubishi Motors Corporation
NEC Corporation
Nihon Kohden Corporation
Nippondenso Company Ltd.
Nissan Shatai Company Ltd.
Omron Corporation
Sony Corporation
Tadano Ltd.
TDK Corporation
Toshiba Corporation
Toshiba Machine Company Ltd.
Yokogawa Electric Corporation
Zexel Corporation

Italy

Alcatel Telettra S.p.A.
Asem S.p.A.
Bull Itailia S.p.A.
C.AR.EL S.r.L.
Calearo S.r.L.
Carraro S.p.A.
Comau S.p.A.

Dataconsyst S.p.A.
Diavia S.p.A.
F.A.I. S.p.A.
Ferroli Industre Riscaldamento
  S.p.A.
Fiam Filter S.p.A.
Gate S.p.A.
I.R.C.A. S.p.A.
IBM Semea S.p.A.
Ina Rullini S.p.A.
Lombardini S.p.A.
Mandelli S.p.A.
Marposs S.p.A.
Marzorati Tecnica Industriale S.p.A.
Otis Italia S.p.A.
Prima Industrie S.p.A.
Riva Calzoni S.p.A.
S.I.L. Met S.p.A.
S.M.A S.p.A.
Saes Getters S.p.A.
SAFOP S.p.A.
SCM S.p.A.
Seima Italiana S.p.A.
Seleco S.p.A.
SGS Thomson S.p.A.
SIT S.p.A.
SIV S.p.A.
Valeo S.p.A.

CHAPTER 2

PATHS OF IMPROVEMENT IN PLANT OPERATIONS

ROBERTO FILIPPINI, ANDREA VINELLI, and CHRIS VOSS

In Chapter 1, Tony Salvatori was considering Six Sigma and other initiatives, such as enterprise resource planning (ERP). Today there are many technological, organizational, and managerial practices and initiatives available for plants that are seeking to improve operations. The choice and the sequence of these is wide, leading to two questions: (1) What are the paths of improvement that plants should take when they adopt initiatives in order to improve operations? (2) What are the results obtained along the various dimensions of performance from these different paths? In this chapter we will offer some answers to these two questions.

In effect Tony Salvatori was facing exactly these questions as he set about to make plant improvements. What path should he take to make improvements over time? Any new initiative, such as Six Sigma or ERP, must consider the initiatives already implemented and should fit with the manufacturing strategy to form a coherent path of innovation.

To realize the desired performance improvements, such as cost, quality, and delivery, the initiatives adopted must be consistent with each other and with the overall aims of the firm. In order to avoid the risk that the organization will reject such initiatives during adoption, this consistency must be checked at various levels. In particular, organizations must strive to achieve consistency between the various production initiatives, even though it has been shown that many improvement practices, when adopted, can create synergies within operations (i.e., JIT [just in time] and MRP [manufacturing resource planning], or quality practices and human resource management, etc.).

Thus, choosing among the many possible alternatives available and then deciding where to start from and which areas to prioritize is a problem of

19

critical importance for plants that have decided to embark on the path to improvement. For plants that have already set out along this path, reviewing their choice, managing the implementation, and managing priorities are all fundamental problems.

Over the past 10 to 15 years, experience has shown that plants very rarely decide to launch several improvement programs simultaneously. This is because most plants must necessarily operate with limited economic, financial, and human resources and because they are well aware of the complexity of managing change, of coordinating different initiatives that lead in more than one direction simultaneously. Thus plants usually decide to focus on a limited set of improvement initiatives, those that they consider to be more important than the others at any one time.

As a result, initiatives for improvement tend to be adopted one after the other in a sequence of implementations that effectively define a *path*. This path may only become apparent as it is followed; it may not necessarily correspond directly to the direction of the first of the initiatives taken by the plant. It is simultaneously the expression of the history of the plant, the decisions made in the past, the future of the plant, and the strategy taken in order to achieve its objectives for the future. The path that a plant follows represents the synthesis of its vision of what being a high performance plant really means.

Although there is a vast amount of material about achieving high performance, in our opinion there is still some confusion about how to get there. First of all, we would argue, as we have in Chapter 1, that there is no one "best way." It is true that the quest for excellence in production usually means adopting many improvement programs, but if they are to be successful, these programs cannot be merely a random selection of the available tools and techniques. Even though each program may be launched correctly, it is unlikely to succeed if the plant simply imitates the actions of its best competitors and has no precise overall strategy that aims to ensure its own success in the future.

The spirit of HPM is, effectively, the search for high levels of production performance through the adoption of an integrated and coherent system of innovative manufacturing practices. Thus, even though there is an overall, common aim shared by all plants that decide to improve their performance, what it means in practice to achieve this aim and how the aim can be achieved (which path to take) remain important questions about which there is, currently, very little implicit or explicit guidance available.

Although the term *improvement* refers both to the initiatives that companies take in relation to various operations and to their results in terms of performance, in this chapter, we use *path of improvement* to mean those patterns of decisions that are related to the initiatives actually taken by a plant. Thus a path of improvement can be seen in terms of the number of initiatives adopted, the type of initiatives adopted, the period in which they were introduced, the level of diffusion of each initiative within the company, and the sequences/priorities accorded such initiatives.

The first area that we will discuss is whether, amidst all the many possible variations that the path of improvement may have, there are consistent patterns of similar paths followed by high performing manufacturers. To be able to identify clusters of plants that take similar paths when adopting initiatives, although they follow independent strategic and decision-making processes, is of great practical value.

We have discovered the common factors and differences among those high performing plants that follow similar paths. The common factors and differences concern the national context, the industry, and the operating and business environments. It is extremely useful for plants, when making decisions about their future strategies, to know that in recent years some improvement paths have been associated with inherently better performance and to know also under which conditions and contexts such routes proved successful.

Eight initiatives have been considered in this empirical international study: (1) flexible automation, (2) design computerization (CAD), (3) supplier relations, (4) JIT, (5) MRP, (6) quality management, (7) human resources, and (8) simultaneous engineering. Details on the definition of these initiatives can be found in the appendix at the end of this chapter.

If one looks closely at these eight initiatives, it is clear that some are principally of a technological nature, such as flexible automation and design computerization, whereas others are mainly organizational, such as human resources, supplier relations, and quality management. Thus it was decided to further classify the initiatives in terms of whether they were *hard-based* or *soft-based*. This is consistent with many other studies that have led to the concepts of *hard* and *soft* being part of the common language of manufacturing management.

Thus, we will speak of *hard-based* initiatives when the intervention presents marked technological/structural features, such as flexible automation and design computerization, and of *soft-based* initiatives when the proposed interventions are primarily of an organizational-managerial/ infrastructure type, such as supplier relations, JIT, quality management, and human resources.

In the case of both MRP and simultaneous engineering, we have chosen a mixed classification because these two initiatives require both changes at the organizational-managerial level and investments in technology and specific hardware and software systems.

IMPROVEMENT INITIATIVES: ADOPTION PATTERNS IN DIFFERENT COUNTRIES

When launching and adopting an improvement initiative, many country-specific variables are important. These include the economic, financial, and industrial situations in a country; the rules and laws that govern the competition; and the cultural and organizational factors of a country that, for example, render a Japanese company very different from an Italian company.

Until quite recently, the country context was, without a doubt, a determining factor for each plant's structure and competencies. However, this difference is gradually being eroded by the globalization of markets and by the consequent need for a plant to operate internationally if it is to survive in the face of global competition. Thus, today, there is greater homogeneity than ever before in both the management and the production practices used by plants operating in different countries. However, this homogenization process is a gradual one and, as yet, only partial. Thus if we look at the past 25 years, both the times and the methods of adoption for the eight initiatives considered here still reflect differences among the plants operating in the five countries analyzed in this study.

Within each area of improvement in operations (initiative), specific areas were analyzed; and for each plant studied, the year in which the single initiative was launched was recorded. In the case of initiatives that included multiple elements, the year of start-up of the first was considered. During data collection, a check was made that the initiatives undertaken by the individual plants had not subsequently been abandoned. Table 2.1 shows the levels of adoption of each initiative studied in the five different countries at the end of 1973, 1983, 1988, 1993, and 1998.[1]

The data in Table 2.1 show that there have been marked differences among the five countries in the timing of the adoption of many of the practices. In particular, in the four soft-based initiatives considered—quality

Table 2.1
Improvement Initiatives Adopted by Year and Country

	Quality Management	Flexible Automation	Supplier Relations	CAD	JIT	MRP	Human Resources	Simultaneous Engineering
Italy								
1973	6%	9%	3%	3%	0%	3%	3%	0%
1983	31	22	6	12	0	34	12	9
1988	69	44	25	72	22	59	37	18
1993	100	75	62	97	62	78	81	41
1998	100	86	85	99	80	84	92	68
United Kingdom								
1973	0	8	8	8	0	8	0	0
1983	8	33	8	33	0	33	0	0
1988	42	50	25	67	8	50	25	8
1993	83	67	75	92	67	75	83	67
1998	96	75	91	98	88	86	98	86
United States								
1973	5	4	0	0	4	8	8	0
1983	12	28	4	32	16	24	12	0
1988	44	68	12	64	40	60	28	8
1993	72	88	56	92	72	80	76	44
1998	85	95	79	98	90	87	90	70
Japan								
1973	62	5	40	2	0	12	52	5
1983	85	50	62	42	50	32	75	17
1988	87	70	65	77	65	50	80	27
1993	88	87	70	90	75	60	82	42
1998	89	94	73	95	83	68	84	55
Germany								
1973	13	10	0	17	0	10	20	0
1983	30	33	0	43	7	40	33	0
1988	40	70	13	73	20	47	43	7
1993	70	90	47	93	43	63	77	40
1998	85	95	75	98	78	75	85	70

management, human resources, JIT, and supplier relations—plants in Japan adopted these much earlier than plants in the other four countries. During the 1960s and 1970s, the West lagged far behind. However, from the mid-1980s on, initiatives involving quality management, supplier relations, JIT, and human resources began to take off rapidly in the West. By the end of 1998, adoption levels were fairly similar in all five countries—a clear sign of the way in which plants in the West had learned about the organizational and managerial methods pioneered in Japan.

The adoption of the soft-based improvement practices in plants with historically very different cultures and traditions is illustrated in Figure 2.1, which shows the relative spread of these soft-based initiatives over time in comparison with the hard-based initiatives. (This was calculated for each country as the percentage of all initiatives launched that were soft-based versus hard-based.) As can be seen, in 1986 the five countries were in very different positions, with Japan having a much higher propensity to adopt soft-based initiatives than countries in the West. This positioning on soft-based versus hard-based can partly be explained by the traditions and the environment of each nation. More specifically, Japan showed a marked propensity to adopt soft-based methods, unlike the Western plants, especially those in the United States and the United Kingdom, whereas Italy appears to have fallen somewhere between the two extremes. Germany, in the 1980s, tended to lean toward hard-based initiatives and does not seem to have emulated the behavior of either the United States or the United Kingdom or of Italy. Plants in these three countries, from the late 1980s on, began a race to catch up on the ground lost to the Japanese plants in terms of the adoption of the soft-based methods that they perceived as being fundamental for the new production models that were being established in that period.

Figure 2.1
Adoption of Soft versus Hard Practices by Country

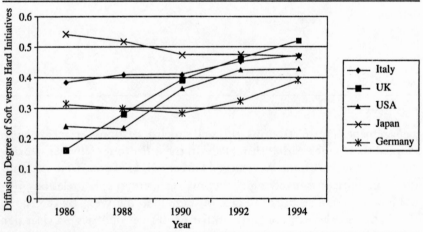

Indeed, the 1994 and the 1998 analyses confirm this trend toward con-vergence. Recently even Germany has shown a marked propensity to adopt soft-based initiatives, and it would seem that all the countries in the study, whatever their historical traditions, are moving toward ever more homo-geneous behavior on all of the eight initiatives under study.

The marked differences between countries in adoption of different methods can be illustrated by two specific areas: quality management and design computerization (see Figures 2.2 and 2.3).

These two are, nonetheless, largely representative of the progress of all the soft-based and hard-based initiatives studied. Both are commonly seen as central to competitiveness. The data in Figure 2.2 confirms what has already been said. It highlights the fact that Japanese plants invested in soft-based initiatives very early on, and shows the later catch-up of West-ern plants that has taken place since the mid-1980s. On the other hand, Figure 2.3 shows design computerization has been adopted at much the same rate in all five countries studied. This shows that although the Japa-nese earned a lot of praise for having placed their trust in soft-based im-provements very early on, this did not stop them from adopting hard-based initiatives as well. This is not true for the European and U.S. plants; al-though they kept pace with their Japanese competitors in the adoption of hard-based methods, they took much longer to introduce the soft-based initiatives.

Figure 2.2
Adoption of Quality Management Practices

Figure 2.3
Adoption of Design Computerization Practices

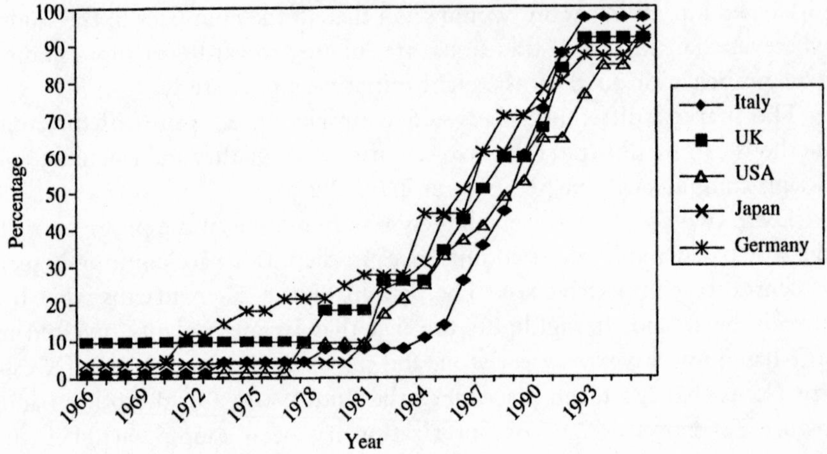

As in the case of a plant's country of origin, the sector in which it operates would also seem to have some degree of influence on the choices made. There are some differences that can be noted from Table 2.2.

The machinery sector appears to be the most "hard-based" sector of all, in that there has been a high percentage of adoptions of both flexible automation and design computerization and, simultaneously, the lowest percentage of initiatives involving JIT, human resources, quality management, and supplier relations. At the opposite extreme comes the auto

Table 2.2
Adoption of Practices by Industry

	Electronics	Machinery	Auto Suppliers
Quality management	87%	78%	85%
Flexible automation	83	85	81
Supplier relations	67	56	60
Design computerization	93	93	89
JIT	65	52	72
MRP	72	74	62
Human resources	76	72	91
Simultaneous engineering	46	33	53

suppliers sector, which appears to be the most "soft-based" sector, in that both JIT and human resource initiatives are most widespread and there has been the lowest rate of adoption of hard-based initiatives. Indeed, the auto sector supply chain was the cradle in which the new lean production model was developed through the introduction of innovative organizational-managerial techniques and practices into JIT production processes, human resources, quality management, and supplier relations, as described in the well-known work by Womack, Jones, and Ross, *The Machine That Changed the World.*[2]

However, as stated earlier, the increasing globalization of competition is gradually eroding such differences and encouraging the transfer and diffusion of managerial and production practices that have proved successful in some sectors to other industries, sectors, and countries.

To conclude, data relating to the adoption of the eight different types of initiatives in the five countries studied have revealed differences, in some cases marked. But the most recent data have shown that for some of the initiatives, there have been high levels of adoption in all countries and sectors. This confirms that the most successful improvement practices are well known and recognized and are being adopted everywhere almost without distinction between sector or country. Thus the differences between plants' behavior when they are seeking to improve their performance must be sought elsewhere. Perhaps the difference might be found in the path they have been following for some time because this path can throw light on the way in which plants' capabilities have evolved, and are evolving, and on the search for elements that differentiate them.

As an Italian plant manager, Tony Salvatori wondered how these different paths might affect his decision making. Rather than seeking "best practices" in any particular situation, he decided to look at his manufacturing strategy and the context of his situation. Based on these factors, a future path of plant migration can be adopted. This would help answer the question of whether to implement Six Sigma and ERP as the next initiative at this time.

IDENTIFYING DIFFERENT PATHS OF IMPROVEMENT

Organizations have a set of choices in the paths of improvement to achieve high performance levels. Given that it is difficult, indeed probably unwise, to have too many initiatives, the first choice is the sequence of initiatives chosen. The second choice relates to the focus of adoption. Companies can choose to adopt just a subset of practices.

To identify the different paths of improvement, we focused on two variables to describe the different paths taken: (1) the number of initiatives adopted as a proportion of the total initiatives considered and (2) the degree of priority accorded to soft-based initiatives as opposed to hard-based initiatives.

The first variable, total number of initiatives, makes it possible to distinguish between two groups of plants. The first group adopts all or almost all the available production practices—whether they be of technological or of organizational-managerial type—with the aim of improving along the greatest possible number of performance dimensions. The second group is more focused and invests only in certain types of initiatives, those they feel are most important for them in their quest for maximum competitiveness. The second variable concerns the type of initiative emphasized—the degree to which soft-based initiatives (organizational-managerial type) or hard-based initiatives (technological type) are emphasized.

Using these two variables—(1) the number of initiatives adopted over the total considered and (2) the soft-based versus hard-based index—*cluster* analysis was carried out.[3] To analyze the clusters we can categorize them based on the two variables. A *Full Adopter* is a plant that has adopted all or almost all of the initiatives considered; a *Selector* is one that has adopted half or less than half of these initiatives. On the other hand, the terms *Soft, Hard,* or *Intermediate* indicate the plant's position regarding the soft-hard index of precedence by indicating the degree to which the plant has prioritized soft-based or hard-based initiatives, or an intermediate position. As is illustrated in Figure 2.4, there are six possible positions.

From the analysis, four distinct groups of plants emerged out of the six possible. They were: (1) Hard Full Adopter—39 plants, (2) Soft Full Adopter—53 plants, (3) Hard Selector—28 plants, and (4) Intermediate Selector—14 plants.

The next step was to review within each cluster the sequence in which the initiatives had been launched by each plant. It was found that plants in the same cluster seem to have followed similar sequences, whereas there was a substantial difference between the common sequence followed across clusters. The sequence in which initiatives had been adopted was termed the *path of improvement.* It shows the specific order, in terms of time, in which a plant had launched each initiative irrespective of the year in which each one had been implemented. The same path could have been followed over

Figure 2.4
Classification of Possible Paths of Improvement

Full Adopter	Hard Full Adopter	*Intermediate Full Adopter*	Soft Full Adopter
Selector	Hard Selector	Intermediate Selector	*Soft Selector*

a short or a long interval of time, before or after, early or late, when compared with other plants in the sample. The paths of improvement identified are shown in Figure 2.5.

The Hard Full Adopters introduced the hard technologies—flexible automation, design computerization, and MRP—before other initiatives. In contrast, the Soft Full Adopters invested much later in this type of initiative and initially implemented organizational-managerial innovations, such as quality management, human resources, and supplier relations. Intermediate Selectors invested in only a limited number of initiatives, but they did so in both soft (quality management) and hard (design computerization) and mixed hard/soft (MRP) initiatives right from the start. The least adopted initiative in all four groups was simultaneous engineering. No Selector plants adopted it, and even the Full Adopter's had so far only introduced it downstream of all other initiatives.

No group emerged that clearly fit the Soft Selector category. Indeed plants that had adopted soft-based initiatives had usually adopted all of those that were included in this study.

There are a number of observations that can be made from this data. First, there are far more Full Adopter than Selector plants. Most of the plants studied—almost 70 percent—adopted all or almost all of the initiatives included

Figure 2.5
Paths Followed by Various Types of Adopters

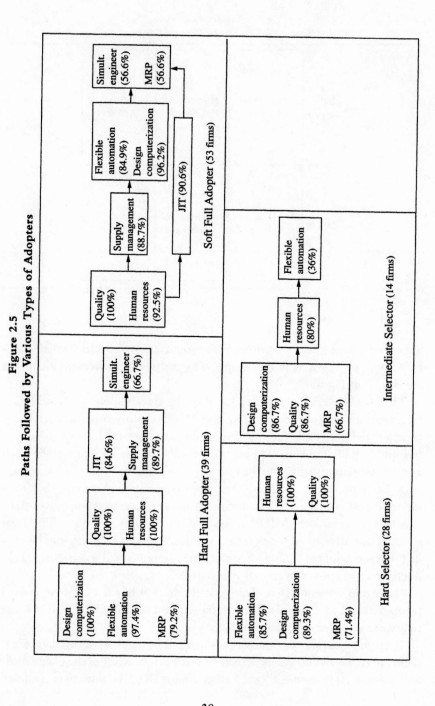

in the study, which indicates that many plants follow the policy of investment in many different areas of practice. Only one-third of the plants decided to adopt a policy of focusing on specific interventions, typically investing in approximately half of the initiatives studied, and only after such interventions were already widespread and well known. Furthermore, exactly half of the plants in the sample initially prioritized technological-type interventions.

In looking at his own plant Tony Salvatori felt that it could be described as a Hard Selector. He believed that the future of his plant should be to continue this path rather than to abandon it. Therefore, he will continue to be selective in implementing new initiatives but will implement those that fit with the existing Hard Selector path. This could include soft initiatives, now that many of the hard initiatives are fully implemented.

ARE DIFFERENT PATHS OF IMPROVEMENT FOUND IN DIFFERENT CONTEXTS?

Plants vary widely, so one important question for managers is whether there are important differences among plants that followed different paths regarding the nature of both their internal and external environments. In order to examine this question, it was decided to consider a set of the variables that would be able to describe the internal and the external environments in which the different plants operated: variables relating to the specificity of the products made, the markets served, and the real production situation of each of the plants studied.

Distinguishing competencies of each plant were examined in order to see whether there was a marked difference among them. The types of competencies considered were those relating to process technology, human resources, and management systems and to external relations with both customers and suppliers. Analysis of the results showed many differences in terms of both production characteristics and specific competencies.

To illustrate the different paths followed by different plants, a brief description of two of the four paths is given, together with short case studies of a plant that followed each path.

• *Hard Selectors* are typically plants that have a small number of product families, fulfill the low number of orders per month, and concentrate on just a few production lines. Production volumes of the main product are not very high. Furthermore, on average, the level of specific competencies is lower than in other groups. These are plants that tend to rely on technical aspects of their products and production in order to obtain competitiveness.

A typical Hard Selector plant in our database produces packaging and pallet transport systems for the food industry. The product depends heavily on a design and a production process that is made to measure for the client. The technical office has a critical role regarding the design of customized products, which are often "one of a kind." The plant's strategy has therefore centered on design computerization (CAD—computer-aided design), linking design and manufacturing (LAN—local area network), databases for components, and automatic generation of the bill of materials (MRP systems). Design computerization as well as the ensuring of efficiency and timeliness in work methods have also made possible the synergistic use of know-how accumulated through successful solving of earlier similar problems. The nonrepetitive environment, low unitary production volumes, decentralization of most of the production, and the use of engineering to order have meant that investment in shop floor automation or in JIT practices is not felt to be appropriate.

• *Soft Full Adopters* are plants that are distinguished by their high volumes of production, which are much higher than those of the Hard Selectors, by the more complex bills of materials (BOMs) of the products made, and by intermediate levels of standardization. The level of distinguishing competencies in these plants is generally high, particularly regarding the competencies required in process technology, management systems, human resources, and external relations, which would seem to distinguish this group from the others.

Another company in our database is a world leader in motorcycle production. The product life cycle is relatively short, the product is seasonal (with 50 percent of sales between May and August), and many of the components used are produced by subcontractors spread out all over the world. The critical performance factor is time to market. The strategy adopted was to achieve flexibility in volume and mix through investments in human resources and suppliers. Continuous training programs were mainly aimed at high and middle company levels (managers and supervisors). Information about company production objectives and results was made available to workers on production lines, many of whom were seasonal; and at the same time, jobs were simplified and standardized. The number of suppliers was reduced; and a raw materials and components warehouse, managed by a firm specializing in logistics, was created outside the factory, downstream of the suppliers. Lines were supplied either directly by external suppliers by Class A codes (by volume or value) or through external transit points by B and C Class codes. In order to avoid holdups in production, the plant invested in Total Quality programs, which were extended not only to all processes within the plant, but also to suppliers both of logistics services and of raw materials and components. Based on the initiatives adopted, this plant was identified as a Soft Full Adopter.

We have now identified four different paths to improvement and highlighted how two of them tend to be used in different plant contexts. Further contextual differences are the country of origin and the sector.

PATHS OF IMPROVEMENT AND PERFORMANCE

Now that alternative improvement paths have been identified, it is important to examine the performance of plants that embark on different paths. Establishing links between an improvement initiative and performance outcome is, perhaps, the most critical and interesting aspect of a study on improvement paths, particularly if it is possible to relate to the context of an individual plant.

To examine the relationship between paths and performance, five performance areas were considered, each closely linked to plant operations: (1) quality, (2) time, (3) cost, (4) flexibility, and (5) inventory turnover. *Quality performance* was based on conformance to standards and was assessed by evaluating the percentage of scrap or rework. For *time performance*, two different dimensions were considered: the delivery lead time and on-time delivery performance. *Performance on costs* was estimated through the relationship between cost of production and total sales. The indicator of *flexibility* was the time horizon adopted to freeze planning, on the basis that a shorter time offers more flexibility. The ratio of production cost to stock levels was taken as the indicator of *inventory turnover*.

Evaluation of a plant's performance cannot ignore either the strategic aims of the plant itself or the specific nature of the critical factors required for that plant's success. In the case of the latter, both the industrial sector to which the plant belongs and the environmental and production context (e.g., country, characteristics of production, volumes, etc.) may play an important role in determining the absolute, or real, level of performance. For example, a two-week delivery time could be considered to be an excellent, a poor, or a completely unacceptable level of performance, depending on the sector, the plant, and the product considered. To reduce distortions of this type, each performance measure was standardized by plant sector. In addition, in order to control the different aims and characteristics of plants, the values obtained in each of the performance dimensions were weighted in relation to the declared strategic aims of the individual plant.[4]

ANALYSIS OF PERFORMANCE LEADERS

To analyze the impact of path choice on performance, we examined the percentage of plants taking each path that could be considered performance leaders. Performance leaders were defined as being in the top 25 percent of the sum of the performance factors, weighted as described earlier. The results are shown in Table 2.3. The numbers in the column give the percentage of plants from the four paths within this top group. More plants following the Soft Full Adopter path have managed to obtain a high all-round level of performance to match their strategic aims. Indeed, one plant in three, in the Soft Full Adopter group, is in the top 25 percent. They are followed, quite closely, by the Hard Selectors, of whom one in four is in

Table 2.3
Leadership in Overall, Weighted
Performance, by Path and by Country

	Percentage of Group That Are Leaders (in the Top 25% 25% Weighted Performance)
Hard selector	25
Intermediate selector	13
Hard full adopter	18
Soft full adopter	34

the top 25 percent. There are relatively few Hard Full Adopters in this top group, and even fewer Intermediate Selectors.

An important result of this analysis, from a practical point of view, is that it has shown that plants following any of the improvement paths are capable of achieving high levels of performance in areas that suit their strategic aims.

Among the highest performing, a large number of *selectors,* those who have specifically selected a set of initiatives rather than fully adopt all the practices, casts some doubt on the "one best way" perspective. In addition, there is not a clear indication that it is better to invest in soft-based initiatives before investing in hard-based initiatives, which also casts doubt on the one best route, as proposed by some.

Returning to Tony Salvatori for a moment, it is clear that he has made the right decision not to simply adopt the next innovation that comes along. Rather, he has carefully considered the path of initiatives that have been adopted, along with his current manufacturing strategy, in an attempt to fit the new initiatives with the existing situation. Because his plant has been a hard selector, his current and future choices will be affected by this path.

CONCLUSIONS AND SOME MANAGERIAL IMPLICATIONS

Over the past few decades, adoption of innovations in production practices has followed a classic S curve associated with the diffusion of innovations. This was found, in different ways, for all the initiatives in all the countries considered. The S curve appears not only for technological type innovations, but also for organizational–managerial type innovations. Usually, if it is to adopt a new initiative before its competitors do, a plant must already possess some special skill or capability. Such choices then force the plant's competitors to imitate, if they are not to fall behind on their performance. However, this process of emulation is neither simple nor straightforward; and the results are not guaranteed. Because they are systemic, hence very complex, the initiatives considered here cannot simply be "acquired" in the same way that a plant would acquire, for example, a new machine. These initiatives demand complex efforts and long time periods for their implementation, and it is here that the differences between each plant's capabilities can prove decisive in determining the success, or failure, of the intervention undertaken. It is almost impossible for most plants to sustain the costs of investing in many areas simultaneously. Plants must first decide what their priorities are and then choose. Furthermore, they must not neglect the most recently available practices, those that are at the beginning of their cycle and only just beginning to be introduced.

The improvement path is not merely a simple sequence of interventions or improvement initiatives implemented by the plant. Rather, it tells the story of the plant's innovations and highlights the evolution of the plant's own capabilities. Most managers accept the notion that the innovations and experiences of a plant are closely interdependent and that the path is dependent on its capabilities.

Empirical analysis from the study has made it possible to identify some elements that differentiate among the choices of improvement initiatives: priorities regarding the type of initiatives (hard versus soft initiatives) and the magnitude of the investment carried out in terms of the number of initiatives adopted. The plants studied followed four main paths: hard- versus soft-based initiatives and many (full adopters) versus few initiatives (selector).

Is there really one path that guarantees success? The answer cannot be, "This is the best path of all," but neither can one say, "It depends entirely on the individual situation." Overall, we found a favorable relationship between interventions taken and performance in the case of the Soft Full Adopters, that is, in the plants that introduced technological innovations "after" they had acted on human resources, quality, and supply management. This confirms the opinion of some who maintain that interventions concerning new technologies will be more successful if the organization (both internal and external) functions within a situation where quality and reliability have already been established. Although the Hard Full Adopters, which concentrated first on technological innovation, are well placed regarding costs, they are not well placed in terms of many other areas of performance, notwithstanding their subsequent efforts to introduce soft technologies. However, introducing only or mainly soft initiatives, but just a limited number of these (Intermediate Selectors), does not result in excellent performance either. Such plants should probably broaden their area of intervention to include technological innovations as well.

Given that, elsewhere in this book, evaluations of the impact of the initiatives studied here on performance will be presented. What other indications regarding "how to choose the right path" have emerged from this study? We have shown how the country of origin, its industrial culture, and, perhaps, reigning fads (i.e., particular interest and popularity of specific innovations at any one time) can all influence a plant's choice of initiatives. Indeed, the influence of the country of origin has often led plants to concentrate their investments in a specific direction. For example, in the past, technological innovations were being prioritized in many European countries, while at the same time, in Japan, many plants were involved in soft initiatives, especially those aimed at improving quality. What can one say about the future? The influence of a country's context can be positive if the plant finds itself in an environment where skills and information about innovations circulate fast and freely. But this is not always the case. Perhaps plants should try to be less influenced by the prevailing local conditions when introducing innovations, especially if they themselves compete at the international level. Thus channels of communication that can gather and circulate experiences gained in different countries and situations should be set up in an attempt to make it possible to recognize, from the outset, both large and small innovations implemented by the best plants and to evaluate both the effectiveness and the potential applicability of these innovations in other contexts in relation to each plant's specific strategies.

In-depth analysis of the plants in this study has shown that innovation per se does not guarantee positive results. Both hard- and soft-based innovations require a great deal of tailoring to a company's needs and context. The specific skills a plant already possesses are, once again, the determining factor in this type of adaptation. Furthermore, innovations that are developed and are successful under specific conditions, not all of which are necessarily known, and under other conditions may well result in unsatisfactory performance. For example, flexible manufacturing systems have been found to be suitable for limited production of a final product that has many variations, while JIT has functioned best in repetitive production situations. CAD/CAM (computer-aided design and manufacturing) technologies seem to have functioned best for plants with complex products, while developing relationships with suppliers that are based on cooperation requires structural and cultural situations that favor partnership.

In other words, it is not simply a problem of choosing innovations but, above all, a need to adapt any innovations to the specific operating context of the plant. This requires a high, consolidated level of knowledge and clarity regarding the plant's ultimate objectives. The fact that it is difficult to imitate innovations, especially soft-based ones, derives both from the complexity of the skills involved in developing and refining the innovation and from the difficulty of learning from the new model. But even hard-based innovations require more than just technical skills, and they too must involve organizational and management competencies if they are to be successful. The fact that all three kinds of competencies—technical, organizational, and managerial, which are based on very different types of knowledge—are called into play simultaneously when new initiatives are introduced into the plant's operations makes it difficult for many plants to imitate those that are already best producers at the international level.

NOTES

1. Some of the 1998 data has been estimated based on trends and a limited sample of data.

2. Womack, J., D. Jones, and D. Ross. *The Machine That Changed the World*, New York: Rawson Associates, 1990.

3. *Cluster analysis* techniques can be used to describe firms on the basis of their path to improvement. The whole sample is divided into groups, or clusters, and these are then analyzed to see whether there are typical paths to improvement that are revealed by the sequence in which each firm has adopted the initiatives within each group. It is also possible to see whether "preferential" pathways can be identified or whether certain paths are characteristic of a specific country or type of firm. After, one can try to discover whether specific paths are associated with better performance when compared with others.

4. A correcting factor was applied to the absolute value of each area performance on the basis of whether the firm itself considered that particular area strategically important (e.g., order winning criteria) or whether the firm saw the performance as an "order qualifier," that is, one necessary in order to compete. For example, if a firm declared that its main objectives were time and flexibility considered these to be the critical factors for its business success—and consequently put less emphasis on cost containment, actual costs performance was weighted less important than it would have been for a firm that had cost leadership among its main objectives.

APPENDIX

DESCRIPTION OF THE EIGHT INITIATIVES

Flexible automation is the attempt to combine the advantages of fixed automation with those offered by programmed automation. Using this method, plants are able to obtain simultaneously low costs per unit and a high degree of flexibility. Flexible automation is defined as an advanced integrated system of hardware and software that makes it possible to design and produce automatically a predefined variety of products. There are various types of flexible automation, four of which were considered in this field study: (1) flexible manufacturing systems (FMS), (2) automated transport and warehousing, (3) production cells and numerical production, and (4) computer numerically controlled (CNC)/direct numerically controlled (DNC) production.

Information systems that support design and production activities fall into the category called *design computerization*. The specific elements examined here are CAD (computer-aided design), CAD/CAM connections (computer-aided design/manufacturing), CAE (computer-aided engineering), CAPP (computer-aided process planning).

Two main types of interventions have been identified regarding the redefinition of *supplier relations*. These are (1) reduction in the total number of suppliers with whom the plant collaborates and (2) the development of partnership relations. Reduction in the number of suppliers is in line with the widespread belief that too many suppliers can prove an impediment for many improvement activities, especially for *total quality management* (TQM).

The *JIT* (just-in-time) approach includes many types of interventions, all of which aim to achieve the production and distribution of the required quantity only, in the exact place, at the precise time, using the minimum of necessary equipment, human resources, and machines. These interventions include both the philosophy of improvement/elimination of waste in production and a set of specific techniques (e.g., reduced *setup* times, product simplification, *pull*-type planning in production, etc.).

Improvement activities associated with *material requirement planning* (MRP) and in *manufacturing resource planning* (MRPII) are grouped under the heading of *MRP* initiatives. The second of these (MRPII) includes not only planning for material requirements but also planning and control systems for the whole firm, for example, capacity plans, stock checks, production costs, financial aspects, and engineering.

Quality management means conformance to specifications and conformance to customer requirements as regards to the product and services offered. The concept of quality has been evolving: in the past quality management was synonymous with inspection, whereas today it has come to have a far wider meaning and has become part of organization, design, production, marketing, and, lastly, the customer. This broader definition, often known as total quality management (TQM), can be pursued through multiple activities (e.g., statistical process control, supplier quality, quality circles, production standardization, product redesign, value analysis, and empowerment).

Initiatives that fall into the *human resources* area are those that are often thought to be the foundation for achieving the plant's quality, efficiency, flexibility, and service objectives. Among the numerous initiatives available, those considered in this empirical study were the level of employee involvement, the existence of work groups, and the dismantling of centralized structures through a new definition of hierarchies.

Simultaneous engineering describes an approach in which the different phases of new product development, from the first basic idea to the moment when the new product finally goes into production, are carried out in parallel. This total approach makes it possible to take into account, right from the first phases, all the characteristics of the product life cycle, quality, costs, and the problems production may encounter. In this way, engineering changes can be carried out very early in the cycle. Thus simultaneous engineering makes it possible both to cut the time required to introduce new products and to drastically reduce design costs, above all, because engineering changes are carried out so very early on.

CHAPTER 3

LINKING PRACTICES TO PLANT PERFORMANCE

MICHIYA MORITA, E. JAMES FLYNN, and PETER MILLING

When a plant implements practices such as small-group problem solving, setup time reduction, and statistical quality control, it is done because they are expected to lead to improved competitive performance. However, managers often pursue such practices as though "the more, the better" was their motto. This chapter has two objectives: (1) to see whether high performing plants are characterized by this general expectation and (2) to explore the processes and structure of practices implemented in high performing plants. Identifying the key practices used by high performing manufacturers is important in making plants everywhere more competitive.

HOW CAN STAGNANT PLANTS COME BACK?

We have all heard stories of how even high performing plants sometimes fall into stagnation. However, the excellence of these plants is demonstrated by the way they are able to return to their growth path. Are there reasons why some plants can achieve dramatic turnarounds, while others falter in similar efforts? Let's consider the example of Honda Motors.

Honda's Resurgence

Honda has been regarded as one of the most innovative firms in the automobile industry. Although it is the most recent entry in the Japanese automobile industry, it has grown quickly, due to its innovative product line, including Civic, Accord, Prelude, and NSX. However, Honda's growth stagnated during the early 1990s. Its products did not appeal to the market,

41

especially the Japanese market. During the same time period, Honda's competitor, Mitsubishi Motors, grew because of its successful entry into the recreational vehicle market, which was the only growth market in the stagnant Japanese automobile industry during that time period.

In order to create an opportunity, Honda's top management designated a team to develop a new recreational vehicle model. The team made trips all over the world, visiting the parking lots of shopping facilities to ask owners how they felt about their vehicles. Based on their market research, as well as on their thoughts as engineers, the team members came up with their concept of the new vehicle. Though Honda's president supported their plan, it was ultimately rejected, due to its high development costs. This was the third rejection of a recreational vehicle model at Honda Motors, all for the same reason.

Even though resources were tight, the team did not give up because they knew Honda would have no opportunity to enter into the recreational vehicle market if they missed this occasion. They modified the original model concept, this time paying more attention to development cost. The first step they took was to fax their concept of the new model to the plants, because they felt the model would not be possible without buy-in from people in the plants. The team asked the plants whether the model was feasible. The answer was that the model was too tall to be assembled on existing lines. Competitors had previously made a broad line of models, including nonpassenger cars, such as buses and vans, so they had no problem with their existing assembly lines. In contrast, Honda had developed a product differentiation strategy, to appeal to the young generation driver with weak brand loyalty for Toyota or Nissan. Most of its models were characterized by low height and a "fun-to-drive" feel. Precisely because of Honda's successful small-car strategy, it did not have the facilities to develop a taller recreational vehicle.

The problem was ultimately solved by cooperative and committed people on the shop floor. They cut the horizontal bar crossing over the assembly line, leaving some length on both ends and raising the remaining portion of the original horizontal bar. This freed enough additional inches of height so that the model could be assembled on the existing line. The new height was acceptable to the design concept and was the maximum the plant could make without significant investment.

However, the recreational vehicle model created other problems in the plant. For example, the model was so long and high that the tail end of the roof hit the ceiling over the paint pool when the body was tipped forward to be dipped into the pool. This time, the designers came up with the solution. They could not shorten the length of the roof or make a protrusion inside by depressing the part of the roof because such treatments would destroy the design and the obstacle-free room that were emphasized in the model concept. Instead, they cut the tail end of the metal roof so that the body could tip forward without hitting the ceiling. They replaced the removed metal part of the roof with a fashionable spoiler. Other changes included development of a new flat suspension and fuel tank system and use of the Accord platform. The total development cost for the revised model was 35 percent less than the original rejected estimate.

This model is called the Odyssey. It is characterized as being spacious (children can easily move about freely inside, including walking through the center without bending over) and having a fun-and-easy-to-drive feel. These and other design features were all developed after considering problems with existing models in the world market. Odyssey turned out to be quite successful, which gave Honda the ability to escape from its stagnation in the Japanese market.

Lessons from Honda

This case illustrates the use of effective linkages between people, functions, professional expertise, and actions. It is important that the design was initiated locally—many people in the Honda organization were against it because Odyssey had an odd shape and was not tall like competitors' vans. The sales and marketing people wanted a clone of the competitors' products, which were selling well. The essence of this success story was the genuinely close cooperation and the devoted efforts of both designers and people on the shop floor. Linkages were extended across all necessary functions in order to develop the car.

Honda successively introduced several new recreational vehicle models in the same way, after the success of Odyssey. Most of them have sold very well, and Honda has come back. Honda is one of the very few Japanese automobile companies that was profitable, even during the Japanese economic stagnation. The success of Odyssey provided an important lesson to Honda in using the leverage of linking necessary functions and people. Honda has rebuilt its business process, from development to sales, with a goal of creating effective linkages throughout the entire organization.

This case also suggests that efforts to create the power of linkages are more likely to be initiated when a company faces some type of crisis. In other words, when things are going well, it's a lot tougher to make the effort to create linkages. In order to sustain the momentum of linkages, a company has to put forth a challenging agenda, coming from higher visions and aspirations. Honda had always been in a scarce resource situation, including people, money, and consumer awareness. The only weapons available to Honda were hard work, the wisdom of its people, and the linkages among them, in addition to Honda's vision to be the champion of car manufacturers. The development of Odyssey reflects Honda's open culture, respecting free discussion and communication, going far beyond the norm. The idea of such communication would not have emerged if the plant had not been equipped with an open culture.

LINKAGES BETWEEN PRACTICES AND PLANT PERFORMANCE

We expect that if people do their jobs well plant performance will improve. However, the relationship between the strength of practices and overall performance is actually determined by how improvements are combined with each other. If, for example, a problem with the defect ratio of shipped products emerges, the focus of line workers may turn to raising the level of quality control practice. But if the real cause of the increasing defect ratio lies in poor product durability, due to the difficulty of fabricating the product, quality control efforts will not be effective in reducing the defect ratio. Instead, efforts of product engineers in redesign of the production fabrication process should be combined with quality control practice.

When we consider the factors involved in these relationships, we find that one important common element exists: the linkages between practices. The relationship between effort and the enhancement of quality practices is often affected by intermediate practices, such as technical assistance or motivating employees. The best way to execute a particular practice may not be obvious to employees, no matter how much they are empowered. Some technical assistance and initiatives may be necessary. In addition, there may be a lack of willingness of people on the shop floor to improve practices. Thus, their motivation may be weak. There is an interaction between technical and behavioral factors. The relationship between practice levels and final performance depends on how the target practice is related to other practices.

In order to achieve better performance, we believe that the linkages between practices should be the strategic target of the plant. Quality can be improved by promoting, for example, consideration of quality problems during the product design stage, quality control activities on the shop floor, the development of human resources to cope with quality problems, an information system that includes quality information and data, the involvement of customers and suppliers in quality design and control, and the involvement of top management in quality. The plant should design linkages among all of these practices in order to increase quality performance. Lack of achievement in any of these practices may cause problems that prevent realization of the plant's goals.

The strength of the linkages depends on how exhaustively relevant practices are involved in the linkage design, as well as on how well each is improved. If the coverage and the integration of the practices are appropriate, the plant will achieve higher competitive performance. If its capability to design and manage the linkages is high, the plant will be able to realign its linkages quickly, even when the competitive situation changes.

Due to the underlying technical and behavioral relationships among practices, linkages among practices improves competitive performance in three patterns: (1) the levered linkage pattern, (2) the trapped linkage pattern, and (3) the transitive linkage pattern.

- *Levered Linkage.* The *levered linkage* pattern emerges once the level of one practice is improved—the level of other practices is also improved. For example, once workers are educated and trained well, small-group problem solving will be enhanced, and the commitment of employees will be raised, improving shop floor capabilities. When the capability of the shop floor improves, barriers to the introduction of new product or process technologies will be lowered. Motivation to strengthen new technology development will also be enhanced.

Behavioral factors work as people are influenced by the conduct of others, an influence that can operate both positively and negatively. If a person is motivated to work in the same direction as his or her friends, it is positive; people are motivated to work hard if their friends do, and vice versa. Negative behavioral factors cause people to behave in the opposite direction. If they see others are not motivated to work, they try to work hard, and vice versa. The ideal behavioral situation exists when a person is motivated to work positively with others when they work hard and to work negatively with others when they work poorly. Because the quality of work is ultimately determined by the commitment of people, management of behavioral factors is critical in realizing the leverage of the linkages.

• *Trapped Linkage.* The *trapped linkage* pattern emerges where a poor practice deters other practices. As in the case of levered linkages, both behavioral and technical factors combine to cause this pattern. For example, normally aggressive research and design (R&D) people may become conservative when they see poor quality control capability on the shop floor. Poor control of suppliers' quality or delivery discourages just-in-time (JIT) practices on the shop floor. This vicious cycle remains intact as long as people's motivation remains the same. Negative behavioral factors provide the means for escape from a trapped linkage structure (i.e., the lack of motivation on the part of some people causes others to work even harder to correct the problem).

• *Transitive Linkage.* In the *transitive* pattern, people may try hard to improve their activities in some practices, however, their efforts only prevail locally. Their efforts are not enough to visibly improve final performance, so they are not recognized by the organization. A plant that tries to improve its poor performance may focus on those practices that appear to be most relevant, such as product quality. However, the plant may find its efforts are not effective, in the short run. Its efforts may begin to wane, unless the plant can make a systematic effort over time. The transitive pattern is perhaps the most critical because it can evolve into either a levered or a trapped pattern, depending on the scope and the integrative nature of the linkages, as well as on the quality of individual efforts.

The average level of practices indicates which pattern prevails in the plant. The levered linkage pattern produces the highest average level of practices, whereas the trapped pattern yields the lowest average level and the transitive pattern falls in between. In a levered linkage pattern, each practice level is high, due to the enhancing momentum of the other practices, causing a virtuous cycle. On the other hand, practices get interactively worse in the trapped linkage pattern. The transitive linkage pattern will have some good and some poor practices simultaneously.

Generally speaking, the levered linkage pattern leads to effective performance in all dimensions of performance (quality, delivery, cost, etc.). Suitable alignment patterns may be different, depending on which competitive performance dimensions are emphasized, but the linkage structure should be established among the practices relevant to the chosen dimensions of performance. Performance will be improved by a levered linkage pattern, if the plant knows how to create a levered linkage.

LINKAGES IN PRACTICE

Parallel Gap in Practices

Figure 3.1 shows the relationship between average performance on the competitive performance measures and the average practice level in the plants in the five countries we studied. It shows that plants with better practices had better performance.

We grouped the practices into nine *drivers:* (1) strategic, (2) technological, (3) managerial, (4) measurement, (5) production system, (6) quality environment, (7) quality control system, (8) shop floor, and (9) human resources development. The drivers consist of functionally similar sets of practices, such as human resources or quality management practices. In Table 3.1 each driver is defined and its component practices listed. Figures 3.2a, 3.2b, and 3.2c compare the average practice levels of the nine drivers of the above- and below-average performers for each industry. We found that drivers in the high performance group are better than those in the low

Figure 3.1
Relationship between Practices and Performance

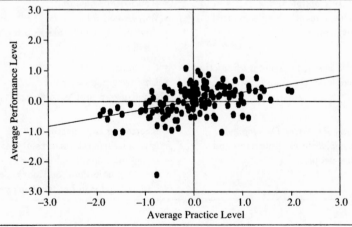

Table 3.1
Drivers and Component Practices

Driver	Component Practices
1. *Strategic:* Degree to which the company implements effective strategic practices.	• Communication of manufacturing strategy • Formal strategic planning • Manufacturing-business strategy linkage
2. *Technological:* Degree of readiness to adopt necessary new technologies and to implement effective new-product introduction.	• Anticipation of new technologies • Interfunctional design efforts • Product design simplicity
3. *Managerial:* How effectively managers motivate employees and coordinate interfunctional decision making.	• Coordination of decision making • Supervisory interaction facilitation • Shop floor contact
4. *Measurement:* Effectiveness with which plant uses measurement systems to grasp the situation, to take appropriate actions.	• Dynamic performance measures • Internal quality information • External information
5. *Production System:* Efficiency of the production system.	• MRP adaptation to JIT • JIT delivery by suppliers • Daily schedule adherence • Equipment layout • Setup time reduction
6. *Quality Environment:* Effectiveness with which customer and supplier involvement is fed into internal quality systems.	• Customer satisfaction • Supplier quality involvement • Customer involvement
7. *Quality Control System:* Degree to which the plant achieves internal quality control.	• Rewards for quality • Process control • Top management leadership for quality • Feedback
8. *Shop Floor:* Effectiveness of practices of people on the shop floor.	• Commitment • Employee suggestions • Small-group problem solving • Maintenance
9. *Human Resources Development:* Means plant uses to develop and deploy people.	• Recruiting and selection • Manufacturing/human resources fit • Multifunctional employees • Documentation of shop floor procedures • Task-related training for employees

Figure 3.2(a)
Practice Levels in the Electronics Industry

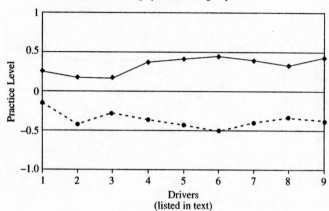

Figure 3.2(b)
Practice Levels in the Machinery Industry

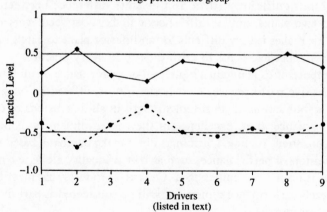

Figure 3.2(c)
Practice Levels in the Transportation Components Industry

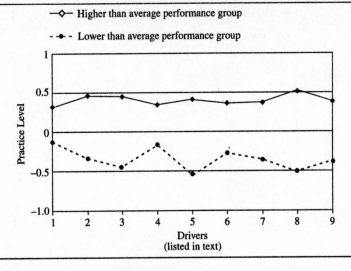

—◇— Higher than average performance group

- ◆ - Lower than average performance group

performance group, across all drivers and in every industry. This shows the way that levered and trapped linkages work on every group of practices in a linked fashion; the differences hold consistently across all practice drivers, which we call a *parallel gap*. The parallel gap defies the idea of skewed or unbalanced excellence; that is, emphasizing isolated practices, such as JIT or quality control, will not lead to overall excellence of the plant. If a particular plant is considered excellent because of its excellence in quality, for example, other practices will also be better than those in the average plant. More significant, if there is a definite difference between performance in two plants, similar differences will exist between every driver in them. This makes it very difficult for an inferior plant to catch up with a superior one, because it has to cross the gap in every practice.

It is important to maintain a high level of performance on all dimensions of competitive performance. Cost leadership or differentiation should be built on a foundation of consistent strength in all dimensions of competitive performance; thus, excellence in the chosen dimension is built on a strong, consistent baseline. Customers don't make purchases based on a single dimension of performance, such as cost or quality alone, even though one dimension may be emphasized. Only the products that meet the standard in each performance dimension will be considered as part of the candidate set for purchase.

We examined differences by country by looking at differences in the perceived need for improvement. Table 3.2 shows that there is a common pattern across the countries. The perceived need for improvement in product performance, quality conformance, on-time delivery, and customer service and support is relatively low in each country. These are all types of performance that can be evaluated directly by the customer. Excellent performance in these characteristics is basic for marketability.

On the other hand, the perceived need for improvement in manufacturing cost, speed of delivery, production cycle time and speed of new product introduction is higher. Both speed and cost performance are relatively invisible to the customer. If they are inferior, however, it will be tough for a plant to compete. Stockpiling inventory in order to deliver on time, low profitability of products, longer planning horizons for production, and delayed introduction of new technologies into products become handicaps to maintaining performance at a competitive level.

We studied the relationship between the improvement requirements of the speed performance and marketability performance sets. The greater the requirements of the marketability performance set, the greater will be the requirement of the speed performance set, and vice versa. The plants that are behind in speed have more difficulty in competing.

Levered Linkages Support the Competitive Foundation

Levered linkages contribute to the foundation for marketability performance. For example, linked behaviors speed various activities. Coordinated linked actions on the floor are required to shorten the production cycle

Table 3.2
Perceived Improvement Requirements by Country

Competitive Measures	Germany	Italy	Japan	United Kingdom	United States	Average
Manufacturing cost	2.56	2.88	2.51	2.85	2.64	2.68
Quality conformance	2.21	2.21	1.77	2.35	1.86	2.08
On-time delivery	2.33	2.50	1.93	2.35	2.28	2.27
Speed of delivery	2.45	2.50	2.17	3.00	2.52	2.52
Production cycle time	2.67	2.94	2.57	2.75	2.52	2.69
Speed of new product introduction	2.67	2.94	2.48	2.90	2.90	2.77
Product performance	2.15	2.12	1.65	2.15	2.25	2.06
Customer service	2.28	2.35	2.26	2.30	1.97	2.23

time. Also, linked actions between R&D and production, through simultaneous engineering, shorten the product development time.

Levered linkages improve all dimensions of performance. As the interactive enhancement of practices diffuses throughout the plant, conflicts between activities, which tend to cause tradeoff relationships, can be avoided. An effort that is narrowly focused on a specific measure is a typical cause of the tradeoff problem. If Honda had not had the constructive interactive relationship between its product development and manufacturing people, Odyssey never would have succeeded, or it would have suffered from a tradeoff, such as high price or low product profile. This sort of tradeoff relationship will soon emerge between the competitive measures in the plants with low levels of linkages.

Effectiveness of Levered Linkage Changes with the Size of the Organization

We found that the stronger the average practice level, the better the average performance. However, we also found a size effect that was analogous to the effect of the number of parts on the reliability of a product. As the number of parts in an item increases, the required quality level for each part also increases in order to achieve the same level of final performance, because the probability of failure of the product is equal to the joint probability of the failure of each of its component parts. The same principle can be applied to management. As more people are involved, the level of each person's practices has to be increased in order to achieve the same level of final performance. As the size of a plant increases, management needs to make extra efforts in educating and coordinating people, as well as in establishing suitable working systems for the employees and management systems, in order to achieve the designed performance. Smaller plants can achieve competitive performance levels at relatively lower levels of their practices. As a plant grows, it will have to improve the level of its practices in order to achieve the same level of competitive performance.

This effort to conquer the law of diminishing returns was pointed out over 40 years ago by Edith Penrose. Many small plants that have grown rapidly have moved into an inferior competitive position because they could not sustain the same level of competitive performance. Peter Senge described this phenomenon as the "lack of systems thinking" syndrome, which creates an imbalance in the plant's activities. In other words, management has failed to grasp the linkage of activities in such plants.

This size effect was confirmed by our analysis, which separated the plants into those with above average and below average practices and performance. In Table 3.3, there are four cells. The plants in Cell 1 are higher than average on both dimensions. Thus, this cell's plants are the high performance and high practice plants. Cell 2 contains overachievers because they have better performance for their low average practice level. Cell 3 contains the underachievers with performance that is poor for their practice level. The plants in Cell 4 are the weakest, with both low practices and low performance.

We found that Cell 1 and Cell 4 plants tend to be larger (in terms of number of employees and sales) than the Cell 2 and Cell 3 plants. Therefore, one implication is that the standard practice level by which a plant's practice level is judged should be lower or higher, depending on the plant's size. The larger the plant, the higher the necessary standard. The plants in Cell 2 are beyond the standard level of performance for their size. Thus, they are not overachievers, literally. On the other hand, the plants in Cell 4 fall short of the standard level of practices and performance for their size. The greater the number of personnel involved, or the larger the business size, the higher the level of practices that is required for them to be competitive. The poor achievers have to improve their average practice level further before they will be in Cell 1.

If the plants in Cell 2 grow in size, they will have to achieve a higher level of practices to even maintain their performance. They cannot remain

Table 3.3
Practices-Performance Matrix

Average Performance Level	Average Practice Level	
	High	Low
High	Cell 1: High Performers Practice level = .64 Performance level = .38 1,811 employees $286,851,000 in sales	Cell 2: Overachievers Practice level = −.30 Performance level = .33 459 employees $87,635,000 in sales
Low	Cell 3: Underachievers Practice level = .79 Performance level = −.44 613 employees $99,308,000 in sales	Cell 4: Low Performers Practice level = −.47 Performance level = −.31 1,123 employees $216,203,000 in sales

at the same level as before. If the standard in their market gets higher as a result of new entries by more competitive firms or existing rivals' additional efforts to improve their practices, it will move the plants in Cell 2 to Cell 4. These plants are more vulnerable to competitive threats because their standard is lower than that of the plants in Cell 1. Enhancement of the standard practice level for a Cell 1 plant is not easy because more practices and people are involved. In a sense, a position as a Cell 1 plant is not easy to sustain because it must keep at least the same high practice level over the entire scope of practices and people. It is easy for the practice level to decline when people reduce their efforts, even in small amounts. Sustaining a high level of all practices requires more energy when more people and activities are involved. This implies that efforts to sustain practices should be systematized. This can include developing codes of conduct, manuals, establishment of formal education systems, mutual learning environments, and plant philosophies that reinforce management systems in Cell 1 plants.

When visiting high performing small plants, we typically found an excellent top manager (management team) who was influential and who managed well. But when it came to relatively large plants, it was their excellent culture that was notable. Small plants tend to put their emphasis on specific aspects that are more critical to their final performance, such as quality control or punctual production. This is due to their top managers' preferences, developed through their experiences and insight, as well as to scarcity of resources, especially talented human resources. Thanks to their competitive position in a niche market where no powerful rivals exist or to their having a strong alliance relationship with more powerful buyers or suppliers, their performance remains satisfactory. Their success depends on how long such a position can last if they remain at the same practice level.

Structure of Levered Linkages

If a plant desires sustained performance, it must be able to cope with any competitive situation. On one occasion, it could be required to make a levered linkage with the floor level practices, while on another occasion, it might need to develop linkages over a more extended part of the process. We can classify two types of levered linkage: (1) *strategic levered linkages* are required to create a strategic thrust, such as the introduction of new products or new processes; and (2) *operational levered linkages* are required to make improvements in existing products or processes.

Both types of levered linkage should be developed in order to maintain high performance over time.

The high performance plant can cope with changing competitive situations if it is equipped with both strategic and operational levered linkages. When a situation requires strategic adaptation, the plant can draw on its strategic linkages. When facing impending competition, it can emphasize its operational linkages. It has high practice levels, on the whole, with both linkages being equally strong.

The ideal structure of the levered linkage in global high performance manufacturers is summarized as a balanced structure between practices, where strategic practices and operational practices are correspondingly high so that the plant can cope with any competitive environment. When the plant adapts strategically, the operational linkage supports the strategic adaptation by improving process management. When it adapts operationally, the strategic linkage assists by institutional and resource provision, such as establishing good cooperative relationships with suppliers and staffing necessary technical personnel and people, as well as by elimination of barriers, such as streamlining organizational structures.

IMPLICATIONS OF THE LEVERED LINKAGE FOR STRATEGIC MANAGEMENT

When a plant tries to function in an uncertain world, factors disturbing its behaviors include uncertainty, a shortage of good ideas, distortions and inconsistent actions due to delay, and differences in focus of the efforts of its people. Variables controlling these factors include time and communication.

Time can mean many things. It can be the elapsed time to do things, such as the production cycle time or product development and introduction time. Its length influences both the timeliness and the cost of actions. Time also determines when people know which problems they must cope with. When they should start their actions is one of the most important determinants of attaining satisfactory results, given their cycle time for actions.

Communication is related to the effects of information. It is a determinant of the quality of information for decision making. People obtain necessary information through various channels of communication. Communication also influences people's motivation. People act and react through communication with others as well as with their organization. Communication changes the state of the human mind or motivation, and it is one of main sources of learning. The learning process consists of accumulating knowledge and increasing the editing capability of the knowledge to create new knowledge for particular problems, as well as creating the understanding of new things. Communication stimulates such a learning process.

When a plant behaves in an uncertain world, strengthening its capability to control time and communication contributes to enhancement of the plants' competence to behave appropriately. The plant must first become aware of problems. Second, it has to create good ideas for action. Developing an appropriate course of action, leading to identification of appropriate performance, is next. The capability to control time and communication improves the working of the process by improving intermediate practices.

The communication system between top managers and the people on the shop floor is implied by the strategic and operational linkages and their interactive relationship. The plant will not take inconsistent actions, such as laying off people who had actually achieved quality improvement or cost reduction because their activities hit the point of saturation or did not improve the overall plant performance directly. Top management of a high performance manufacturer knows when and how their achievements will pay off and how other achievements should be combined with them.

The levered linkage structure assumes mutually responsive and cooperative behaviors among relevant people. Such behaviors will reduce the time required to rework unsatisfactory results or to make modifications due to poor adjustments between actions or people. People can get to know the needs of others through good communication and can grasp what should be done, based on the knowledge of their needs. Such behaviors invoke effective learning.

Many plants enjoy strong performance for a period of time, but it is the *continuity* of strong performance that is a challenge. Plants have to create leverage by linking their activities. For example, practices related to JIT or to quality management may be well known, but sometimes their validity has been argued. This is appropriate when such practices have been implemented in isolation. Local efforts are the litmus test of whether a plant is equipped with appropriate logic for creating strategic and operational linkages and integrating the linkages throughout all practices.

PART II

SPECIFIC HPM PRACTICES

CHAPTER 4

MANUFACTURING STRATEGY: BUILDING CAPABILITY FOR DYNAMIC MARKETS

KIM BATES, KATE BLACKMON,
E. JAMES FLYNN, and CHRIS VOSS

STRATEGIC MANUFACTURING TODAY

Manufacturing strategy has taken an increasingly important role for manufacturing firms. Process technology, turbulent product markets, and close partnerships with customers and suppliers demand significant changes. A business unit's strategies are becoming increasingly visible to competitors, to customers, and to investors through the explosion of information available via print and electronic media and through global participation in capital markets. This has greatly increased the pace with which business units must devise and implement strategies in product markets.

Emerging economies have produced a host of low-cost competitors that are able to challenge even the best firms in the developed economies, increasing the pressure to improve their cost positions. Similarly, constant advances in process technology have increased the need for business units to improve manufacturing processes to keep pace with competitors. Manufacturing managers must be able to combine constant improvement of existing manufacturing processes with judicious investment in new processes, utilizing both human and capital resources in order to maintain their competitive position in product markets. Therefore, manufacturing strategy not only must cope with the rapid change generated by hypercompetitive markets, but also must serve as a resource that is linked to both business strategy and organizational processes.

This constant pressure on manufacturing firms for change and adaptation implies that they must be able to implement strategies quickly and must plan to adapt manufacturing processes on an ongoing basis. The constant need for manufacturing strategy decisions to develop processes, facilities, and human resources has never been more important. However, there has been very little concrete information available for managers about what is meant by *implementing* manufacturing strategy. In our study, three factors emerged in plants that consistently develop and implement manufacturing strategy:

1. Creation of routines within the plant for aligning manufacturing and investment decisions with the firm's business strategy.
2. Plant personnel who know—and communicate widely—how manufacturing supports the firm's strategy in the marketplace.
3. A consistent adoption of manufacturing process innovations spanning decades.

In this chapter, we discuss manufacturing strategy, highlighting the factors that emerged from our study. Next, we compare manufacturing strategy in the five countries studied and explore the relationships between manufacturing strategy and manufacturing process innovation, focusing on how successful plants approached the most important manufacturing process innovations of recent years. Finally, we look at the future of manufacturing strategy in supporting high performance manufacturing, focusing on long-term investments in manufacturing processes through the consistent adoption of manufacturing process innovations.

MANUFACTURING STRATEGY

Manufacturing strategy is the process companies use (1) to build the resources and the capabilities to create competitive advantages and (2) to align their competitive priorities with the marketing function. Firms must choose the most appropriate processes to create and sustain competitive advantage in the marketplace; to develop long- and short-term goals for responding to the increasing challenges in product markets; and to acquire resources in the factor markets for labor, process technology, and information technology. Whereas previous accounts of manufacturing strategy have emphasized a top-down process of aligning manufacturing with top management, and with marketing, our study shows something different.

What is clear from our study is that challenges from competitors, globalization, and technology-driven changes in society have made the strategic role of manufacturing more difficult to define and to implement—and, more important, to maintain long-term high performance. Interviews with successful manufacturing managers in our study showed that they possessed a relentless desire to ensure that their operations possessed the depth, the resources, and the talent to stay on top. Analysis of our performance data suggests that successful manufacturing plants have a history of pursuing appropriate process innovations that support the business unit's strategies for its products and of communicating the strategic impact of decisions widely within the plant.

Because this is one of the first comprehensive studies of manufacturing in five countries with a history of manufacturing excellence, we were able to analyze how manufacturing managers interpret issues related to manufacturing strategy and to sort out a consistent, global interpretation of manufacturing strategy implementation. *Manufacturing strategy implementation* means widely communicating how decisions made within plants relate to business unit goals. We interpret these results as an important message for firms that depend on manufacturing for competitive strength in the marketplace: strategic issues must permeate decision making within the plants. Furthermore, plants with high levels of manufacturing strategy implementation consistently outperformed those that did not, and these plants were the ones that implemented more manufacturing process innovations and adopted them ahead of other plants in their industries.

We believe that this ability to consistently innovate manufacturing processes is a result of a firm's long-term commitment to build manufacturing capabilities and resources that not only align with business strategy but also serve as the source of its evolving competitive advantage. It represents an imperative that manufacturing develop, adapt, and implement new processes that expand the firm's competitive capabilities. One case in our study illustrates this decades-spanning commitment to understanding the strategic impact of decisions, the importance of innovation, and the evolution of capabilities.

Tennant Corporation is a U.S. manufacturer of large floor sweepers and industrial cleaning equipment. It exemplifies a long-term commitment to manufacturing excellence and the ability to implement new technologies. Tennant was one of the first U.S. firms to embrace total quality management (TQM). In the late 1970s, management discovered that Japanese customers had higher expectations than Tennant's North American customers. Japanese customers observed that when Tennant industrial sweepers were not in service, they occasionally leaked oil onto the floor where they were parked. Management realized that if Japanese customers demanded sweepers that did not leak oil when not in use, other customers would quickly develop the same standards, particularly if Japanese competitors gained a toehold in North American markets. Further, if Tennant could not meet these demands, they might lose the Japanese market to competitors. By embracing TQM early, Tennant reinforced its leadership in product features with higher levels of product performance, kept costs low, and was able to maintain its market share.

Several years later, during an economic downturn, Tennant experienced high levels of inventory and took across-the-board cuts in hours and pay in order to maintain its skilled management and labor force. At the same time, management became aware that just-in-time (JIT) practices were enabling firms with higher volume and more repetitive manufacturing processes to create higher product variety and to reduce inventory levels. Management studied JIT and implemented it on a limited basis, but it concluded that most JIT methods were unsuited to its low- and medium-volume processes. However, some JIT practices were implemented in its parts fabrication, and inventory levels were substantially reduced. Nonetheless, mixed-model production eluded management in the early 1980s.

Tennant continued to pursue excellence in manufacturing throughout the 1980s, and by the 1990s it installed a visual, electronic point-and-click information system on the shop floor that allowed its larger models to be manufactured with a mixed-model production process on its four production lines. The information system was an interactive electronic standards manual located at each workstation that enabled workers to check procedures using a graphical user interface. Operators accustomed to one model could refresh their memory on another, reinforcing cross training and speeding the introduction of mixed-model assembly in this medium-volume assembly environment. Workers could click on a picture of the sweeper being manufactured to see a visual representation of the assembly process for a particular component. Standards could be updated centrally, and all operators could be immediately alerted to changes in procedures as they were introduced. Tennant's operators and process engineers were pushed to keep improving production processes.

Ironically, what most excited the process engineers was that they had an information system that was accepted by operators, who were busy lining up new ideas for using the system to improve performance. They had been able to justify the installation of the system of the interactive standards manual because it enabled mixed-model production, but they were most interested in seeing how the technology could be adapted to new projects. Both operators and engineers were able to use this innovation to produce further innovation quickly and effectively by enhancing communication about process improvement. One of the first firms in the United States to adopt this technology in the early 1990s, Tennant used it to achieve a goal first conceived almost a decade prior.

Tennant demonstrated a long-term commitment to achieving strategic goals, through clear communication of strategic goals, and to an environment conducive to innovation. When management first encountered the emerging practice of JIT, they realized that lowered inventories and mixed-model assembly were important strategic goals for competing in their market place, but that the pull system was not suited to a low-volume, complex product. Yet management remained aware of the advantages of mixed-model production and seized an alternate method to achieve it. Throughout the 1980s, management discussed alternative means of achieving this goal in a cost-effective manner, while at the same time implementing several other manufacturing improvement projects.

ALIGNING MANUFACTURING WITH BUSINESS STRATEGIES

Previous discussions of manufacturing strategy claim it is a translation of the firm's product market strategies into manufacturing terms, implying specific guidelines for tactical planning. However, in our study what seems to make the difference, worldwide, is whether firms can implement the right manufacturing strategy at the right time—adapting new innovations to their processes and consistently seeking to differentiate their processes from competitors. Manufacturing management must understand the implications of decisions on their ability to create and to sustain competitive advantage. Aligning manufacturing with business strategies requires expertise in the processes managed, and in the ways that technology developments affect those processes, as well as an understanding of strategic initiatives in product markets.

Our research in manufacturing plants echoes recent research on corporate strategies that emphasizes the need for dialogue between middle managers and top management, creating a strategic direction from the top while middle managers within the plants chart the specific course of investments in process improvements that will best implement strategies. We believe that manufacturing strategy should be a guide rather than a strict blueprint for plant management because changes in process technology and high rates of continuous improvements by competitors create opportunities for manufacturing managers to improve the competitive position of the firm. This is why manufacturing must see itself as a resource and a foundation for the future, not as a reactive instrument to dictates from top management.

Due to the rapid pace of globalization, turbulent hypercompetitive markets are very likely to continue to dominate manufacturing industries. Globalization spreads new ideas for improving manufacturing processes and creates relentless pressure for established firms to match the cost positions of competitors from developing countries. Therefore, high performance manufacturers will have to adopt and implement operations process innovations frequently and rapidly. Plants must achieve excellence in multiple dimensions of manufacturing performance; but, more important, they must have the capacity to rapidly *become* excellent along any dimension of manufacturing performance that the competitive situation demands. Firms with a better understanding of how manufacturing supports and builds business strategy and with confidence that manufacturing contributes to the success of the firm will be able to quickly implement appropriate manufacturing process innovations earlier and more successfully.

We focused on five dimensions of manufacturing strategy: (1) anticipation of new technologies, (2) manufacturing-business strategy linkage, (3) formal strategic planning, (4) communication of manufacturing strategy, and (5) manufacturing strategy strength. These dimensions emerged as a consistent, tightly related set of variables across the five countries in the study (see Table 4.1). They describe important features of manufacturing strategy implementation, from the link between the plant and the business unit, and the use of planning for long-term vision, to the way that strategies are used to guide decisions in the plant. Our focus was on how strategies influence decisions and routines within the plant, as well as on the use of strategies for long-term technology acquisition and configurations of

Table 4.1
Correlations* between Manufacturing Strategy Scales

Scales	Anticipation of Technologies	Communication of Strategy	Formal Planning	Business Strategy Linkage	Strategy Strength
Anticipation of technologies					
Communication of strategy	.52				
Formal planning	.45	.49			
Business strategy linkage	.67	.52	.64		
Strategy strength	.65	.47	.59	.69	

* All correlations are statistically significant.

manufacturing processes. Because of the close relationships between these dimensions, we will refer to them collectively throughout the rest of the chapter, confident that they form a consistent set of dimensions across the five countries studied here.

Nevertheless, there were some differences among the countries in how manufacturing strategy is implemented (see Table 4.2). Germany emerged as the most consistently focused on a strategic approach to manufacturing— with the highest scores on three of the five scales and virtually tied with Japan once. German plants were particularly strong on the manufacturing-business strategy linkage, anticipation of new technologies, communication of manufacturing strategy, and overall manufacturing strategy strength. The Japanese plants scored high on formal strategic planning and communication of manufacturing strategy. These results are not surprising, given the historical strength of these two manufacturing powerhouses. These results are also consistent with the percentage of plants in each country that have adopted formal manufacturing strategy programs, as shown in Figure 4.1. The figure also shows that the German plants are the most recent adopters of manufacturing strategy programs, perhaps accounting for their strong focus on a strategic approach to manufacturing decisions. Not surprisingly, the Japanese plants were early adopters of formal manufacturing strategy programs, and a majority of those plants use manufacturing strategy programs to guide decision making.

Overall, the Japanese and the Germans placed the most importance on manufacturing strategy implementation, followed by the United States and the United Kingdom, with Italy consistently placing less emphasis on manufacturing strategy implementation (see Figure 4.2). The Italian plants tend to be smaller, and they may have less of a plantwide focus on manufacturing

Table 4.2
Manufacturing Strategy Implementation by Country

Manufacturing Strategy Scales	United States	Germany	Italy	Japan	United Kingdom
Anticipation of new technologies	3.47	**3.68**	3.38	3.51	3.33
Manufacturing-business strategy linkage	3.61	**3.95**	3.57	3.27	3.73
Formal strategic planning	3.55	3.67	3.09	**3.84**	3.46
Communication of manufacturing strategy	3.50	**3.65**	2.94	**3.68**	3.38
Manufacturing strategy strength	3.43	**3.64**	3.48	3.55	3.42

Note: The numbers in the table indicate the strength of implementation of each practice on a scale of 1 to 5, with 5 being the highest. The bold numbers show statistically significant differences.

Figure 4.1
Percentage of Manufacturing Strategy Adoption by Year

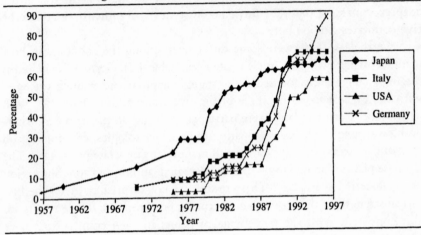

strategy to guide decisions because a small group of managers take responsibility for strategy implementation.

If we look at the contents of the manufacturing strategy questions in our survey, some interesting features emerge. Manufacturing managers must understand not only the capabilities of their own processes, but also the ways

Figure 4.2
Current Manufacturing Strategy Implementation by Country

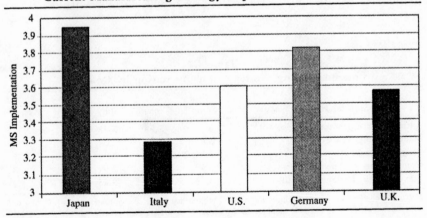

to build and sustain them; and they must understand how strategic change will affect the manufacturing processes they manage. This requirement of manufacturing managers is in marked contrast to previous definitions of manufacturing strategy because it focuses on understanding the impact of strategic changes on the capabilities of manufacturing systems, rather than on the need to be able to communicate to top management, who often are unfamiliar with process technologies. The important point is that both are present in our data, suggesting that the debate about bottom-up or top-down strategic approaches obscures the importance of widespread communication and consistent uses of strategy in decision-making routines. We therefore emphasize the need for manufacturing managers to take responsibility for assessing changes in process technology and the implications of emerging best practices in their industries, along with linkages to business strategy.

The principles of new manufacturing innovations spread like wildfire in industries subject to global competition, through the agency of trade organizations and consulting firms. Aligning decision making requires exploring the possibilities inherent in the innovation du jour—but not jumping on every bandwagon. Manufacturing management must evaluate how new innovations will impact manufacturing process capabilities, whether those capabilities are necessary for achieving strategic goals in product markets, and whether they form a platform for adopting future innovations. Without careful consideration of the impact on operating systems, valuable resources could be wasted on inappropriate changes to manufacturing processes, or opportunities for future innovations could be missed. Aligning manufacturing with business strategies means that manufacturing management must understand the need to view manufacturing from the perspective of its cumulative capabilities.

Toshiba Machine, of Japan, illustrates the importance of manufacturing management knowing and understanding business strategy. Toshiba competes in an industry—wafer fabrication machinery—that has been increasingly driven by economies of scale in production, which puts more pressure on sustaining a low-cost competitive position, particularly for firms in countries with a relative disadvantage in labor costs, such as Japan. In addition, increasingly short product life cycles and complex production processes drive competition in this industry. Therefore, the capability to develop, adopt, and implement the next generation of process technology is a strategic imperative.

The wafer fabrication industry was plagued by overcapacity in the previous generation of production, in which many firms suffered punishing losses due to dropping prices. In the mid–1990s many firms were faced with the necessity of collaborating with competitors and subcontracting excess capacity in order to justify astronomical expenditures on facilities and processes. As a supplier of machinery, Toshiba Machine had to be one step ahead of its customers in developing ever more complex production equipment and in facilitating partnerships with customers in order to successfully produce equipment for this market. The ability to successfully partner with customers means a high degree of interaction between technical personnel at both firms in a labor market that has been ravaged by a sluggish domestic economy and by rapid obsolescence of technical expertise among older, generalist engineers.

In order to compete in this market, Toshiba management had to break with two, traditional, Japanese human resource policies that were part of the policy of lifelong employment with the company. The first policy was that only recent graduates in engineering could be hired. Yet the recession in Japan in the 1990s meant that many talented engineers were available— engineers who had expertise that Toshiba Machine needed and who would be thrilled to work for the company. The second policy was that its engineers had to be generalists—familiar with multiple products. Toshiba had to allow its technical experts to become specialists. Encouraging technical expertise in one area was crucial for supporting short product development cycles in this brutally competitive market. Both decisions have forced Toshiba Machine to adjust to new practices, yet dedication to the goal of creating the next generation of machinery and the high stakes involved in rapid, successful product design have enabled them to adjust. Without a clear knowledge of changes in their product market that favored collaboration among customers, and ever shortening product life cycles, Toshiba Machine might have been unable to build the technical expertise necessary to win orders for the new generation of wafer fabrication machinery.

Because building and exploiting manufacturing capabilities are critical to successful implementation, manufacturing management must be aware of developments in the markets for technology, labor, and manufacturing practices. Management must understand how to renew manufacturing processes in response to market pressures and, more important, how to create and sustain new capabilities. As specialists in production processes, manufacturing management is in the best position to develop, adapt, and understand the implications of emerging practices and technologies. Without an awareness of emerging practices, manufacturing will likely strive to meet performance targets that are no longer relevant, and the organization as a whole will become unaware that it is slipping behind. Manufacturing managers must possess a good working knowledge of the firm's strategies and an understanding of how manufacturing decisions are aligned with business strategies in order to create production processes that excel on multiple dimensions of performance. In addition to and essential to long-term success, manufacturing managers must maintain the ability to develop new capabilities quickly.

COMMUNICATING MANUFACTURING STRATEGY WITHIN THE PLANT

Communicating manufacturing strategies within the plant is just as important as aligning manufacturing decisions with business unit goals. We believe that the dynamic nature of competitive environments makes implementing manufacturing process innovations quickly and effectively crucial for high performance manufacturing (HPM). The dynamic nature is also the reason that "communicating manufacturing strategy widely within manufacturing" emerged as such an important factor in our study. Among top-performing plants, knowledge of emerging technologies is likely to be widely possessed by the managers and the engineers within the plant, making communication of manufacturing strategy crucial to renewing the plant's processes over time.

We also measured the adoption of 11 manufacturing process innovations, creating a 50-year record of adoptions across the five countries (see Table 4.3). Plants that scored high on our manufacturing strategy scales tended to be the plants that had adopted more innovations than their competitors had and had adopted them earlier than competitors did for a majority of the innovations studied. This provides solid evidence that a strategic approach to manufacturing decisions is associated with solid results in building capabilities in manufacturing processes over a period of decades.

In our study, high performance manufacturers were among the earliest to adopt emerging manufacturing innovations, constantly striving to keep production processes ahead of competitors and to anticipate market demands for new achievements. Thus they create and sustain a set of capabilities that fosters not only innovative developments and adaptations but also speedy and efficient implementation of the innovations. The plants that were able to use and implement new process innovations were those plants where strategies were communicated widely. For example, at Citizen Mechatronic in Japan, all personnel, including the plant receptionist, meet with the plant manager monthly to contribute ideas for achieving strategic goals. Although examples of continuous improvement within

Table 4.3
Manufacturing Strategy Implementation and Innovation Adoption

Manufacturing Innovation	Early Adopter Implement More Extensively?	Difference between Adopters and Nonadopters in Manufacturing Strategy Implementation?
Formal manufacturing strategy program	Yes	No
Computer-aided design/ computer-aided manufacturing (CAD/CAM)	Yes	No
Cells	Yes	No
Computer-integrated manufacturing (CIM)	No	Yes: Adopters implement manufacturing strategy more.
Employee involvement	Yes	No
Equipment upgrades	Yes	No
JIT	Yes	Yes: Adopters implement manufacturing strategy more.
Reduction in levels of management	No	Yes: Adopters implement manufacturing strategy more.
TQM	Yes	No
Supplier partnerships	Yes	Yes: Adopters implement manufacturing strategy more.

Japanese plants have become commonplace, Citizen Mechatronic is somewhat unique. The plant manager was a turnaround artist who had been recalled from overseas to save this plant from closure—a scenario much more commonly reported in other countries in our study.

Successfully implementing innovations over the course of decades requires that management understand the implications of the innovation, implement it quickly and successfully, and then explore how the innovation can contribute to achieving new levels of performance. Innovation requires communication. Successful plants readily adapt and implement the next innovation, even as the new innovation makes the most recent one obsolete.

Unless strategic goals are widely communicated within plants, and unless personnel are consistently made aware of how decisions support business unit strategies, plants will be unable to recognize the need to question long-standing practices that prevent it from reaching its goals.

MANUFACTURING STRATEGY ACROSS COUNTRIES

Although we found many similarities among successful plants across countries, we found differences among countries as well. The country of plant location emerged as the most important factor for predicting which manufacturing process innovations would be adopted, and in what order. Clearly plants look to firms within their own countries to gauge the competitiveness of their operations and to glean information about new strategies and manufacturing processes. The country was consistently more important than the industry in predicting the adoption of new innovations. We believe this fact points to the importance of communication links among firms within countries, whether through customers, suppliers, competitors, or consultants. For example, several factories in Japan reported adopting a strategic approach to manufacturing in the 1950s, whereas the earliest that plants in other countries adopted a manufacturing strategy approach to decision making was in the late 1960s (see Figure 4.1). Some of these dates coincide with these firms' postwar revitalization and reflect the importance of manufacturing excellence among export-oriented Japanese firms. These dates also reveal the long-term benefits of a strategic approach to manufacturing decisions. By the mid-1980s, the strength of a strategic approach to manufacturing had leveled off in Japan, with around 65 percent reporting a manufacturing strategy initiative. Early manufacturing strategy initiatives in the other countries began in the 1970s and leveled off in the late 1980s in Italy and the United States. However, factories in Germany were increasingly beginning to embrace a strategic approach to manufacturing well into the 1980s, with almost 90 percent reporting a manufacturing strategy initiative in the late 1990s. What is important is that these differences in manufacturing strategy implementation today are reflected in the plant's pattern of adopting new manufacturing process innovations and that other plants within the country are the examples that seem to be used.

These findings of a linkage among manufacturing strategy implementation, country, and long-term investments in manufacturing process innovation raise some serious issues for plants competing in global industries. Plants must look beyond their own spheres of influence and the comfortable network of competitors within their own countries to foreign firms within their industries if they expect to survive in today's highly competitive global marketplace. Our data show that the number of innovations and the rate at which innovations are adopted have increased over the past decades, which provides empirical evidence of the impact of hypercompetitive environments on plants. We believe that in the future, high performing manufacturers will be required to look both within and beyond their countries to keep abreast of new strategies and new practices within manufacturing.

FUTURE FOCUS ON MANUFACTURING STRATEGY

Our cases and our analysis of performance data across the five countries in our study illustrate the importance of long-term commitment and a strategic approach to manufacturing. Unlike earlier studies, we found that a consistent pattern of strategic initiatives to improve manufacturing processes was necessary. We believe that this finding points to a greater role for manufacturing managers in assessing developments in the factor markets for technology and labor and that analysis of competitors strengths should include plant managers who are familiar with manufacturing processes. In other words, manufacturing strategy also means seizing the opportunities inherent in emerging practices.

This future focus by plant managers became evident through interviews with managers of successful plants. We found that in order to successfully implement manufacturing strategies, manufacturing management must be aware of developments in manufacturing process innovations as they emerge and must have the good judgment to understand which aspects of new practices are appropriate for their processes. Further, management must remain aware of the challenges the firm faces in product markets in order to anticipate which dimensions of performance will become paramount. The successful firms in our study maintained this focus over a decade, and this long-term consistency paid off in the long term, even when conditions in their industry were grim. They did not simply grab onto practices that emerged within their industry but looked across industries to understand new practices. In order to maintain focus and disciplined decision making, firms must communicate business strategy widely and accurately, yet they must avoid a narrow focus on the innovative practice du jour.

CHAPTER 5

HUMAN RESOURCE
MANAGEMENT PRACTICES

KELLY A. MOLLICA, ANEIL K. MISHRA, and BARBARA B. FLYNN

People are the most important source of competitive advantage in virtually any organization. While physical resources, technology, products, and processes can be duplicated across plants and even across countries, people cannot. How people are managed can have a greater impact on the financial performance of a plant than can being in the "right" industry or country, coming up with the "right" strategy, or having access to the most cutting-edge technology. In his book *The Human Equation*, Jeffrey Pfeffer provides compelling evidence across several industries and in different countries that plant profitability is connected to how people are managed, concluding that strong management of human resources yields returns on the order of 30 to 50 percent. Thus, it is appropriate to view human resources as an asset.

Despite such evidence, many organizations still view human resources (HR) as a variable cost. Why is this a problem? When people are viewed as costs, they often become the focus of cost-cutting efforts. For example, management's response to increasing competition and tighter profit margins may be to cut wages, to hire less-skilled workers, or to lay off a portion of the workforce. On the other hand, when human resources are viewed as assets, management will view them as an investment and will expend time and financial resources on management and employment practices, regardless of economic conditions.

Investing in human resources is especially important in high performance manufacturing (HPM) plants, where workforce skills, abilities, and commitment to the organization play a key role in achieving high levels of performance and where the role of knowledge work is becoming increasingly more important.

In the global manufacturing environment, the use of different practices in different environments is also an issue. Because of differences in legal and political systems, it is expected that the use of certain practices will be stronger in some countries than in others. In addition, unique national cultures will influence the attractiveness of various practices to inhabitants of different countries. Thus, we will look at the role of differences among countries in the use of human resource management practices in order to determine the extent of adaptation of practices to differences in legal, political, and cultural systems in the five countries we studied.

HUMAN RESOURCE MANAGEMENT AS A SYSTEM OF PRACTICES

Human resource management is a system of practices and policies designed to influence employees' attitudes, behaviors, and performance. Individual performance depends on having the necessary skills and abilities for the job, plus having the motivation to apply those skills and abilities. Human resource management plays a central role in ensuring that employees possess the needed skills and abilities through, for example, selective hiring and comprehensive training. Human resource management also plays a role in influencing employee motivation through, for example, both monetary and nonmonetary rewards and employee involvement.

Although the use of a single human resource management practice can influence plant performance, the use of a system of human resource management practices is far more effective in influencing the performance of a plant. Strong human resource management practices provide employees with the information, skills, incentives, and responsibility to make decisions that are essential for building a competitive advantage through increased employee competence and motivation, positive work-related attitudes, higher productivity, lower turnover, and development of a progressive company image.

Key human resource management practices include selective hiring, incentive compensation, performance management systems, training and development, sharing of knowledge, and employee involvement. Human resource practices that are commonly associated with high performance are the following:*

---

* Adapted from R.A. Noe, J.R. Hollenbeck, B. Gerhart, and P.M. Wright, *Human Resource Management: Gaining a Competitive Advantage* (Boston: Irwin/McGraw-Hill, 2000).

- Employee selection is intensive and selective.
- Employee rewards and compensation relate to financial performance and other key objectives, such as quality and customer service.
- On-going training is emphasized, and employees are cross trained.
- Employees receive formal performance feedback and are actively involved in the performance improvement process.
- Teams are used to perform work and to make improvements.
- Employees participate in planning changes in equipment, layout, and work methods.
- Employees understand the strategy and the goals of the company.
- Employees understand how their job contributes to the finished product or service.

Perhaps the most important component of a human resource management system is *employee selection*. HPM requires a very intensive approach to the selection of who will become a member of the plant. Selection involves several steps: (1) identifying critical knowledge, skills, abilities, and attitudes that are needed for a job; (2) generating a pool of applicants; and (3) using valid selection tools and methods for evaluating candidates. High performance manufacturers expend a great deal of resources on exhaustive hiring processes. Managers in such organizations take the selection process seriously, as they would any significant investment, rather than seeing it as an unnecessary expense and a waste of time.

Compensation is closely tied to selection because a plant can be more selective if it offers an attractive compensation package. Pay sends a signal to employees about whether they are valued and also about the behaviors and performance that are valued by the plant. It is inconsistent for a plant to claim it is competing on the basis of people while at the same time paying substandard wages. A simple rule of thumb is that people do what they believe will be rewarded. Compensation should be used as a tool to motivate employees and to influence behaviors that affect cost, quality, customer satisfaction, and other plant goals. Two recent trends in compensation frequently used by high performance manufacturers are (1) making a greater percentage of employees' pay contingent on their performance and (2) giving employees a greater stake in the organization through profit sharing and stock ownership.

Training is closely linked to competitive advantage of a plant because high performance manufacturers rely on front-line employee skill and initiative to identify and resolve problems and to take responsibility for quality. Although training often focuses on the basic skills needed to perform a job, it should also focus on broader skills and the knowledge needed to create intellectual capital. Advanced training includes the use of technology to share information with other employees, understanding manufacturing strategy, and knowledge of customer needs. In addition, cross training ensures that team members understand each other's jobs so that they can fill in as needed.

Another important HR area of high performance manufacturing is *performance management*. Performance management can be used to evaluate and to motivate both individual and team performance, focusing employees' attention on plant goals. Ideally, performance management encompasses setting specific objectives, providing timely rewards that are tied to these objectives, and including a feedback and developmental component that encourages coaching rather than punishing. Thus, performance management can tie the interests of a plant with its employees. It should emphasize involving employees in improving their performance by giving them accurate information about those aspects of job performance over which they have control, such as measures of productivity, costs, quality, and customer satisfaction.

Information sharing is an important element in high performance manufacturing because it allows employees, or teams of employees, to make decisions about how to perform their jobs to meet goals and because it shows them how their jobs contribute to the finished product or service. Regardless of how well trained and motivated employees are, they cannot work toward improving performance if they do not have information on the key dimensions of performance as measured by the plant or the training to interpret and use that information.

The *use of work teams* has resulted in well-documented benefits for many plants. Teams increase employee involvement by decentralizing decision making and giving employees greater control over planning and coordinating their work. Some teams are almost entirely self-managed—they perform functions traditionally done by management, such as employee selection, performing inspection and quality control activities, and coordinating directly with customers. Self-managed teams increase employees' sense of accountability and responsibility for the success of the plant.

The Link between Human Resource Management and Performance

Recent research has shown the link between strong human resource management practices and plant performance. The measurement of performance varies across these studies, but performance indicators generally include profit, productivity, quality, efficiency, and customer satisfaction. We briefly highlight some findings here:

- A frequently cited study of 968 firms across 35 industries found that companies that rated higher on the use of human resource management practices had significantly higher profits, sales, and market value and lower employee turnover in comparison to companies with low ratings.
- A joint venture between General Motors and Toyota (NUMMI), in which strong human resource management practices were adopted, produced substantially higher productivity and quality in comparison to General Motors plants using traditional management practices.
- A study of 62 automobile assembly plants in various companies across 17 countries found that product quality and productivity were much higher in plants using a flexible production system with an emphasis on teams, employee involvement, cross training, and performance-dependent compensation.
- A study of 97 manufacturing plants in the metalworking industry demonstrated a link between the use of "human-capital enhancing" human resource strategies (selection, training, performance management, and compensation) and employee productivity, efficiency, and customer satisfaction. This link was particularly strong for companies that competed on the basis of quality.
- Productivity and profits were significantly higher in 30 steel plants that used high performance work practices (careful selection, cross training, and the use of problem-solving techniques) as compared to other plants. Another study of 30 steel plants produced similar results. Plants that used a "commitment" rather than "control" approach to managing employees experienced lower scrap rates and higher productivity.
- The use of "modular" systems (self-directed teams, cross training, and team incentives) by companies in the apparel industry was associated with higher gross margins as a percent of sales, higher sales growth, and lower inventory, in comparison with companies using traditional management practices.
- A study of 15 semiconductor fabrication plants found that management practices emphasizing decentralized decision making, knowledge sharing, team-based rewards, and team training were associated with lower defects and higher productivity.

• Profits were substantially higher and maintenance expenses lower among oil refineries that had built a committed workforce through knowledge sharing, employee involvement in improving performance, and investments in cross training.

Relationship between Human Resource Management Practices and Performance

Table 5.1 summarizes what we learned about the relationship between human resource management practices and plant performance in our study of manufacturing plants in five countries. The plus signs indicate the presence of a statistically significant and positive relationship between specific practices and measures of plant performance. A minus sign indicates a negative relationship; for example, the minus sign in the first line indicates that greater use of individual merit pay was associated with *lower* levels of product quality.

Our findings indicate that there is a core of human resource management practices that is consistently related to plant performance, regardless of which measure of plant performance is considered. These HR practices include:

Table 5.1
Relationship between Human Resource Management Practices and Performance

Practices	Performance Measures					
	Organizational Commitment	Unit Cost	Product Quality	On-Time Delivery	Inventory Turnover	Cycle Time
Individual merit pay			−			
Incentives for group performance		+		+		
Profit sharing		+	+			
Small-group problem solving			+	+		
Cross training				+	+	+
Multifunction employees	+	+	+	+	+	+
Implementation of employee suggestions	+	+	+	+	+	+
Recruiting and selection	+	+	+	+	+	+
Communication of strategy	+	+	+	+	+	+
Manufacturing/HR fit	+	+	+	+	+	
Feedback	+	+	+	+	+	+
Rewards for quality	+	+	+	+		+

- Implementation of employee suggestions.
- Selection of job candidates based on their desire and ability to work in a team, their problem-solving aptitude, and their work values and attitudes.
- Management communication of strategies, goals, and objectives to all employees.
- Performance measures that are related to plant objectives and that are adjusted when goals change.
- Providing feedback to employees on their performance related to quality, productivity, and other types of performance.

This core of practices provides the foundation for high performance manufacturing. These practices are tightly linked with each other. For example, multifunctional employees will be able to provide better suggestions for improvement because of their knowledge and experience with multiple processes. Receiving useful feedback about their performance will better position employees to receive rewards for quality. Job candidates who are hired based on their desire and ability to work in a team are more likely to receive rewards for quality.

In addition to this core of human resource management practices, other practices are related to specific types of performance. For example, providing incentives for group performance is effective in encouraging problem-solving activities targeted at specific goals, such as reduced unit cost and improved on-time performance. Profit sharing encourages a focus on cost reduction, combined with quality improvement. Cross training is particularly effective in improving measures of time performance, including on-time delivery, inventory turnover, and cycle time reduction. These are all measures of just-in-time (JIT) performance, where the ability to cover all positions on the line is critical, due to the lack of inventory buffers. Thus, the first five practices listed in Table 5.1 are performance-specific practices, while the remaining seven practices form part of the infrastructure for high performance manufacturing.

One of the more interesting findings of our study is the lack of a relationship between organizational commitment and pay practices. It appears that employees are not likely to form an attachment to their organization based on the forms of compensation that they receive. In contrast, having input into the performance improvement process, receiving feedback on performance, having knowledge of the strategic direction of the company, and receiving cross training do have a significant and positive relationship on employees' commitment to their organization.

NATIONAL DIFFERENCES IN HUMAN RESOURCE MANAGEMENT

Use of the practices we have described varies among countries because of differences in cultural, legal, educational, and political environments. In this section, we briefly describe examples of some of these differences, discuss cultural influences on human resource management practices, and present findings from our study of practices in five countries.

Differences among countries exist with respect to the most important criteria used in making selection decisions. For example, in hiring new employees, Japanese plants put heavy emphasis on subjective and demographic criteria, such as social skills, age, and gender. German companies, on the other hand, tend to heavily emphasize formal educational credentials.

Employee training varies among countries, partly due to the extent to which there is government involvement and laws mandating training. Germany has a well-structured apprenticeship system, with an emphasis on vocational training, that is regulated by the government. German companies also conduct extensive in-house training. Government involvement in employee training is more moderate in the United Kingdom where plants often upgrade employee skills and competencies by hiring trained personnel from the external labor market rather than by investing in the training of existing employees. There are no statutory training requirements in either Japan or the United States; however, training, particularly cross training, is pervasive in Japanese plants, whereas the extent and the type of training is quite variable across plants in the United States. A 1999 *Industry Week* survey of U.S. manufacturing plants found that about one-quarter provided less than 8 hours of formal training annually per employee, whereas only about 11 percent provided more than 40 hours of annual training per employee.

Compensation and rewards systems also vary greatly across countries, particularly in terms of their emphasis on performance versus nonperformance factors. Compensation in Japanese plants has traditionally been based largely on seniority and age. However, there is a growing trend toward the use of merit pay, and profit sharing is widely used in Japanese plants. Italian and United Kingdom plants generally put little emphasis on compensation based on performance—employees are compensated primarily based on time worked. In contrast, compensation in German plants is heavily based on formal educational credentials, and performance-based pay is not widely used.

Most Western European countries have legislation requiring the representative participation of employees, which is commonly carried out through the use of works councils that have specific legal rights. However, there is variance across European countries in the extent to which work councils are actively used. In many European organizations, representation in works councils is merely symbolic and does not necessarily spread to the shop floor; thus, it cannot be said that true employee involvement is prevalent. For example, the majority of large plants in Germany have active works councils in place, whereas the use of works councils in Italy is inconsistent. In contrast, in Japan, small teams of workers address production issues, and management shares information to help team members make decisions; however, such teams are not fully autonomous. Legal forms of employee involvement are relatively rare in the United States, as compared with other countries. In reality, the use of truly autonomous, self-managed teams and a high degree of employee involvement is not pervasive in most countries, at least not in larger organizations.

CULTURAL INFLUENCES ON HUMAN RESOURCE MANAGEMENT

What causes these striking differences in practices between countries? Certainly, legal and political differences have a great deal of impact. At a deeper level, however, many of these differences are caused by profound differences in national culture. The *culture* of a country is defined as a pattern of assumptions about solving problems that is shared by the inhabitants of a country. Approaches that have been used effectively in the past are passed on to younger inhabitants as the correct way to perceive, think, and feel about similar problems. Thus, the culture of a country develops from years, or even centuries, of success in solving problems in certain ways, so that it becomes unthinkable to consider solving them in a different way.

Geert Hofstede studied the cultures of many countries and developed four key dimensions of national culture that differentiate between countries.

1. *Individualism/collectivism* describes the degree to which people in a country are oriented toward acting as individuals versus acting as part of a group. In countries with individualistic national cultures, such as the United States and Canada, people are expected to act according to their own interests. Their relationship to their organization is viewed as a business transaction; so they believe poor performance or better pay are valid reasons for terminating a relationship, and individual bonuses and incentives are effective motivators. In contrast, in countries with collectivist national cultures, such as many South American countries and Taiwan, people act according to the interests of their group. They see their relationship to an organization as much like their relationship with their families; although performance may determine the particular tasks to which an employee is assigned, it could never be the reason for termination, any more than a child would be dismissed from a family for poor performance. Group relationships are extremely important; thus, group incentives and bonuses are more effective in countries with collectivist national cultures.

2. *Power distance* is the extent to which employees expect that power is distributed equally. In countries that are low in power distance, such as the Scandinavian countries and the United Kingdom, there is limited dependence of subordinates on their bosses and a preference for consultation and boss-subordinate interaction. Organizations in countries that are low in power distance are decentralized, with a flat hierarchy and a limited number of supervisors. The range of salaries between the top and the bottom of an organization is relatively small. In contrast, in countries that are high in power distance, such as the Arab countries and India, bosses and subordinates are considered to be fundamentally unequal. Power is centralized as much as possible, with a large number of supervisory personnel and tall hierarchies. There is a wide gap in salaries between employees at the top and those at the bottom of an organization, and superiors are believed by all to be entitled to special privileges.

3. *Uncertainty avoidance* refers to the degree to which people in a country prefer structured, rather than unstructured, situations. National cultures that are higher in uncertainty avoidance, such as Greece and Switzerland, have a strong need for rules, which leads to punctuality and a talent for precision, illustrated by Swiss watches. On the other hand, national cultures that are lower in uncertainty avoidance, such as the United States and Hong Kong, have an abhorrence of formal rules, establishing them only when absolutely necessary.

4. *Masculinity/femininity* is a dated term used to describe the extent to which aggressiveness and success (versus concern for relationships) are valued in a national culture. A national culture that is high in masculinity, such as that of Japan or Austria, values high earnings, advancement, and challenging work. A "good" (masculine) manager is an assertive, decisive, aggressive, lonely decision maker who considers the facts when solving problems.

In contrast, a national culture that is high in femininity, such as that of Sweden or Thailand, places a high value on good working relationships with direct superiors and on working with people who cooperate well with one another. The preferred mode for resolving conflicts in these countries is compromise and negotiation, and the ideal job provides opportunities for mutual help and social contacts. A "good" (feminine) manager is a less visible leader who is facilitative, rather than decisive, and seeks group consensus.

Hofstede's dimensions of national culture support the idea that there is not a cookie cutter approach for high performance manufacturing. The inhabitants of different countries will feel more comfortable using practices that are consistent with their national culture and will avoid or change practices that are not consistent with their national culture.

Hofstede devised a scale to rate countries on each of his dimensions of national culture. These ratings provide a basis for developing expectations about which human resource management practices will be most effective in each country. In Figure 5.1, we used Hofstede's ratings to develop profiles of the countries that we studied. It shows that the United States and the United Kingdom are quite similar in national culture—high in masculinity and individualism, while low in uncertainty avoidance and power distance. This suggests that plants in these countries are populated with individuals who value challenge, advancement, and aggressive decision making. They are likely to be somewhat creative and may feel constrained by structures for decision making and standardized procedures. These aggressive decision makers would feel free to challenge their bosses and may relish an argument to help cement their opinions. In contrast, the Japanese national culture combines masculinity with collectivism; thus we would expect to see groups, rather than individuals, who value challenge, advancement, and aggressive decision making. The Japanese culture is also very high in uncertainty avoidance, suggesting that the groups would thrive in highly structured decision-making situations. Although the groups are aggressive, they still respect and defer to their superiors.

Figure 5.1
Profiles of National Culture, Based on Hofstede's Dimensions

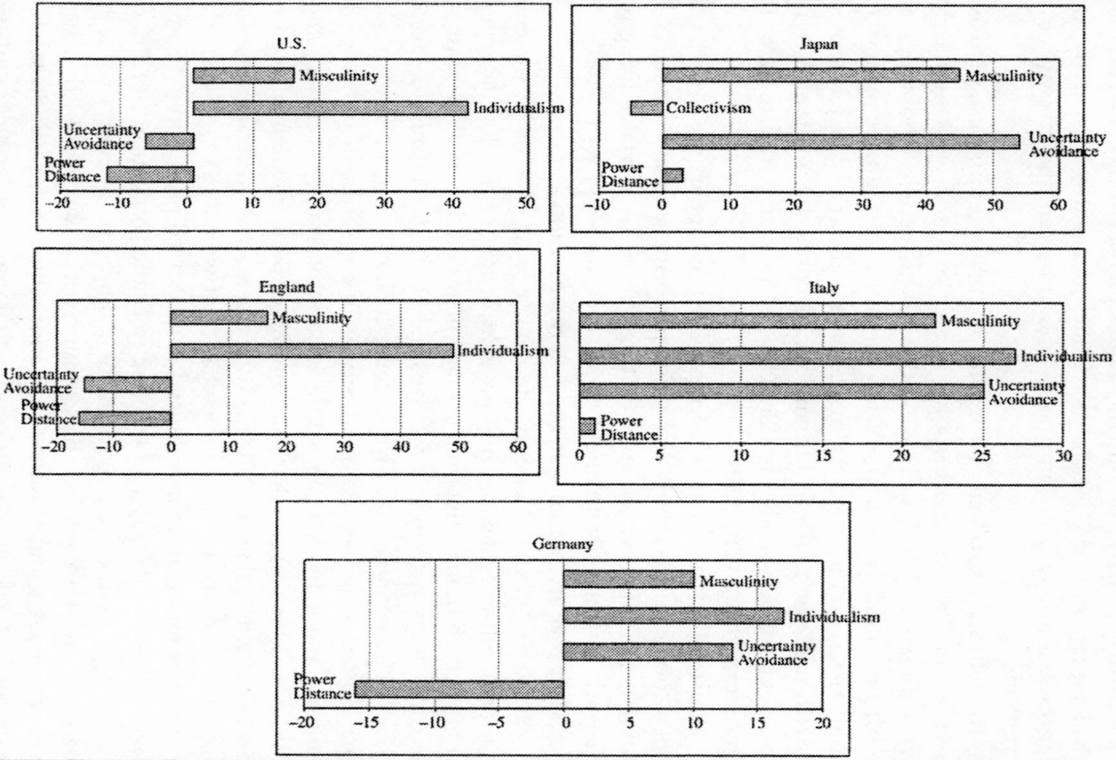

The national culture in Italy combines the aggressive individual decision makers (like those in the United States and the United Kingdom) with a strong preference for structured decision making and deference to authority (as in Japan). We would expect Italian decision makers to work best as individuals, preferring highly structured decision-making processes. The national culture in Germany falls between the extremes on most dimensions. In power distance, the German culture is similar to that of the United States and the United Kingdom, with a belief in the fundamental equality of people, regardless of the jobs they are holding. We would expect this to lead to the use of human resource management practices that support this fundamental equality, such as profit sharing. The German national culture also falls between the extremes in uncertainty avoidance, indicating employees who like the security of rules but who are willing to take some chances as well. It is lowest in individualism, except for Japan, which is more collectivist. We would expect German plants to be inclined to use a combination of individual and group-based approaches, leaning more toward individual approaches. German culture is lowest in masculinity/femininity, however, still on the masculine side, indicating a preference for a mix of approaches; we would expect German employees to be somewhat driven and aggressive but to also feel comfortable with facilitation, negotiation, and consensus approaches.

Differences in Human Resource Management Practices by Country

In our examination of differences in human resource management practices by country, we expected differences among the five countries, based on their cultures, legal and economic structures, and historical precedents. Figure 5.2 reports the results of the first part of this analysis, which focuses on the measures of human resource management practices that were expressed as percentages. For example, 82.4 percent of the Italian plants used individual merit pay, while only 13 percent of the Japanese plants used this approach.

Several findings are noteworthy. First, the use of individual merit pay is lowest in Japanese plants and greatest in Italian plants. Incentives for group performance and profit sharing are used most prevalently in plants in Japan, whereas profit sharing is almost nonexistent in Italian plants. These findings are perhaps not too surprising, given that Japan has the most collectivist national culture of these five countries and that Italy has both a relatively

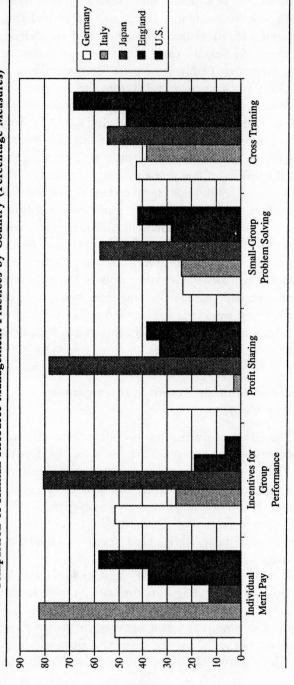

Figure 5.2
Comparison of Human Resource Management Practices by Country (Percentage Measures)

86

high level of individualism and high power distance (the degree to which inequality is acceptable in society). Because of the strong power distance in the Italian national culture, the idea of profit sharing could seem strange to employees and managers in Italian plants, who would perceive managers and lower-level employees as fundamentally different. The relatively high use of small-group problem solving in Japanese plants, compared with plants in the other four countries, reflects Japan's strongly collectivist national culture.

Less striking differences existed among the five countries for the other high performance work practices examined in our study. Most of these are shown in Figure 5.3, which shows our findings for the practices that were measured using employee perceptions about practices on a scale of 1 to 5. The use of cross training of employees and multifunction employees was quite similar across all of the countries except Italy, whose plants reported lower utilization levels of these practices. Indeed, the Italian plants had lower utilization of all of the nonpay, human resource management practices than plants from the other four countries in our sample had. Italy has a higher power distance than Germany, the United Kingdom, and the United States; and higher power distance implies greater dependence of subordinates on bosses and belief in the fundamental inequality of people in different levels of the organizational hierarchy. The national culture of Italy is also characterized by a greater degree of individualism than that of either Japan or Germany. The combination of high power distance and relatively high individualism may influence Italian plants to rely less on human resource management practices that involve group-based incentives or that involve the sharing of information and decision-making power between superiors and subordinates.

We found that the plants in Japan were the highest on most of these practices, followed closely by the plants in Germany and the United States. These were the plants that tended to have the highest scores on most of the measures of plant performance, supporting the link between core human resource management practices and plant performance. The plants in the United Kingdom tended to rank a close fourth on most human resource management practices, while the plants in Italy were a distant fifth. Plants in Italy seem to place less emphasis on human resource management practices than do the plants in the other countries that we studied. However, this may be at least partly due to the effect of size. Because the plants in Italy tend to be so much smaller than the plants in the other countries, there may be less need for explicit human resource management practices.

Figure 5.3
Comparison of Human Resource Management Practices by Country (Scale Measures)

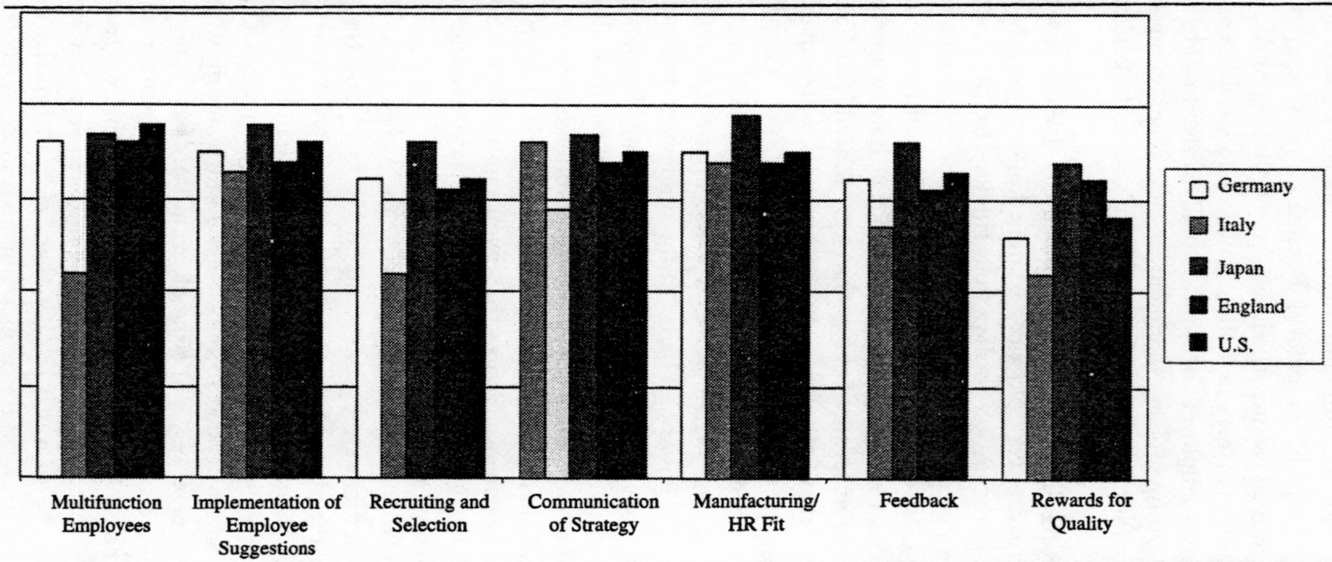

Legend:
☐ Germany
▨ Italy
▣ Japan
■ England
■ U.S.

Categories (left to right):
- Multifunction Employees
- Implementation of Employee Suggestions
- Recruiting and Selection
- Communication of Strategy
- Manufacturing/HR Fit
- Feedback
- Rewards for Quality

CONCLUSIONS

In this chapter, we described human resource management as a system of practices, provided evidence for the link between high performance work practices and plant performance, and presented our findings about human resource practices and plant performance from our study of manufacturing plants across five countries. We explicitly view human resources as assets and investments, rather than as costs. Our underlying assumption is that improved plant performance is achieved through people, who are the most important source of competitive advantage for plants because they cannot be easily duplicated. However, we do not suggest that human resource management practices are the *only* factor that affects plant performance; the best performance will be achieved by the plants that have a strong infrastructure of high performance manufacturing practices.

Overall, the use of human resource management practices was associated with improved performance in each of the five countries we studied, although there was variation across countries in both the use of particular human resource practices and their link with performance. This supports the idea that a system of human resource practices can significantly influence plant performance, but that the particular practices that are most effective in a given environment will vary.

It is important to consider the moderating effects of plant strategy, structure, technology, and other differences among plants. The link between human resource management and plant performance is dependent on how well the human resource practices that are implemented by a plant fit its strategic goals, such as cost, quality, or flexibility. A particular strategy requires unique knowledge, skills, abilities, attitudes, and behaviors that are reinforced and supported by human resource practices. For example, plants that pursue a quality strategy focus on product reliability and customer service, which are heavily dependent on employees' technical and problem-solving skills and on their ability to work in teams. In order to achieve these competencies, plants should engage in selective hiring to attract people with the necessary skills and abilities, provide comprehensive training, and implement a reward system that reinforces its quality objectives. These same practices might have less impact on performance in a plant largely pursuing a cost strategy, in which strategic objectives are achieved through routinization of work, which requires less discretion on the part of the employees.

We also emphasize the importance of linkages among human resource practices. Human resource practices must support each other in order to have an influence on employee behaviors or performance. For example, an emphasis on teams and group problem solving would not be well supported by individual pay practices. Cross training is of little use if employees return to their jobs and are not given the opportunity to apply what they have learned in different jobs or are not rewarded for doing so. Performance management systems must be based on measurement criteria that reflect the key objectives of the plant and that focus only on those behaviors and outcomes that contribute to these objectives. They should also be consistent with the compensation practices of the plant.

We caution against a "one-size-fits-all" approach to the implementation of human resource management practices in the pursuit of high performance manufacturing. As we discussed earlier, societal and cultural norms and traditions vary greatly across different countries, making it difficult to assert that a particular set of human resource practices will be equally effective from one culture to another. U.S. plants that expand into a multinational arena quickly learn that a generic approach to managing people often does not work.

On the other hand, human resource practices can sometimes be adopted cross-culturally with success. For example, despite considerable differences between U.S. and Japanese national cultures, U.S. plants that adopt Japanese-style human resource management practice are often as productive as Japanese plants. This suggests that the performance-enhancing advantages of Japanese plants can be attributed to human resource management practices rather than to cultural differences. These practices have a strong influence on the organizational culture that exists at specific plants. A strong organizational culture can override the influence of national culture; for example, the use of Japanese-style human resource management practices can override the tendency toward individualism that is typical of plants in the United States. As employees of these plants use group-based practices and observe their effectiveness, the organizational culture will move toward being more collectivist, even in a strongly individualistic national culture, such as that in the United States.

Thus, the influence of culture on practices and performance is complex. In general, we recommend adapting human resource management practices to fit unique national cultures in order to achieve the best performance. Our findings indicate that there are several alternative human resource management paths to high performance. However, we also found compelling evidence that there is a strong infrastructure of core human resource management practices that is related to high performance manufacturing.

Adopting this group of practices can lead to changes in a plant's organizational culture that will allow it to be effective, even though the practices may appear to be somewhat contradictory to elements of the plant's national culture. Thus, dealing with issues of national and organizational culture implies a process of understanding, adaptation, and capitalizing on the interactions between national and organizational culture.

REFERENCES

Cascio, W.F. *Managing Human Resources: Productivity, Quality of Work Life, Profits.* Boston: Irwin/McGraw-Hill, 1998.

Greer, C.R. *Strategy and Human Resources.* Englewood Cliffs, NJ: Prentice Hall, 1995.

Guest, D.E. 1999. "Human Resource Management and Performance: A Review and Research Agenda." In *Strategic Human Resource Management,* R.S. Schuler and S.E. Jackson, eds., 177–190. Oxford: Blackwell Business, 1999.

Hernan, P. "The Untrained, Unempowered Masses." *Industry Week* 248, 22 (October 18, 1999): 94, 96.

Hofstede, G. *Culture's Consequences: International Differences in Work-Related Values.* Beverly Hills, CA: Sage Publications, 1980.

Hofstede, G. *Cultures and Organizations: Software of the Mind.* London: McGraw-Hill, 1991.

Hofstede, G., and M.H. Bond. "The Confucius Connection: From Cultural Roots to Economic Growth." *Organizational Dynamics* 16: 4–21.

Hofstede, G., B. Neuijen, D.D. Ohayv, and G. Sanders. "Measuring Organizational Cultures." *Administrative Science Quarterly* 35: 286–316.

Huselid, M.A. "The Impact of Human Resource Management Practices on Turnover, Productivity, and Corporate Financial Performance." *Academy of Management Journal* 38: 635–672.

Ichniowski, C., and K. Shaw. "The Effects of Human Resource Management Systems on Economic Performance: An International Comparison of U.S. and Japanese Plants." *Management Science* 45: 704–721.

Kleiman, L.S. *Human Resource Management: A Tool for Competitive Advantage.* St. Paul, MN: West Publishing, 1997.

Meyer, J.P., and N.J. Allen. *Commitment in the Workplace: Theory, Research, and Application.* Thousand Oaks, CA: Sage Publications, 1997.

Noe, R.A., J.R. Hollenbeck, B. Gerhard, and P.M. Wright. *Human Resource Management: Gaining a Competitive Advantage.* Boston: Irwin/McGraw Hill, 2000.

Pfeffer, J. *Competitive Advantage through People: Unleashing the Power of the Workforce.* Boston: Harvard Business School Press, 1994.

Pfeffer, J. *The Human Equation: Building Profits by Putting People First.* Boston: Harvard Business School Press, 1998.

Schein, E.H. *Organizational Culture and Leadership.* San Francisco, CA: Jossey-Bass, 1992.

Youndt, M.A., S.A. Snell, J.W. Dean, and D.P. Lepak. "Human Resource Management, Manufacturing Strategy, and Firm Performance." *Academy of Management Journal* 39 (1996): 836–866.

CHAPTER 6

COMPETITIVE PRODUCT AND PROCESS TECHNOLOGY

FRANK H. MAIER and ROGER G. SCHROEDER

IMPORTANCE OF TECHNOLOGY: THE COMMON SENSE VIEW

Continuous technological change—the process of creative destruction—is commonly seen as a prerequisite for the competitiveness and survivability of companies and whole economies. In the late 1950s, Robert Solow (1957), the 1987 Nobel laureate, stated that 87.5 percent of the increase in productivity can be explained through technological change and only 12.5 percent through economic growth. Michael Porter (1985) characterized technological change as "a great equalizer, eroding the competitive advantage of even well-entrenched firms and propelling others to the forefront," which therefore influences the structure of whole industries. However, Porter also noted that technology has no value for its own sake; being the technological leader in an industry or just being a competitor in a technologically intensive industry is no guarantee of profitability and success.

Nevertheless, the question arises whether outstanding technological performance really is a critical factor for the success of manufacturing. This seems to be true especially in capital-intensive industries, like the machinery, electronics, and automotive suppliers industries that are the subject of this book. Effective implementation and use of technology is cited as a strategic weapon in a company's battles against competition so often that everybody believes in it. But is this really true? Does high performance manufacturing (HPM) require remarkable use and effective implementation of technology? What distinguishes high performance manufacturers from traditional plants, with respect to technology? What technology practices and dimensions are high performance manufacturers using? And what are the implications for management? These questions are addressed here.

93

FACETS OF TECHNOLOGY

Technology is a multifaceted term and, hence, should be clarified first. Traditionally, technology comprises the plant's products—product technology—and manufacturing processes—process technology or manufacturing technology. Nowadays, a third aspect comes into play and gains more and more importance: information technology.

Product Technology

Often technology is thought of only as product technology. In this view, "technology" only comprises the kind of technology included in a company's products and in what makes a company's products work to meet customer needs. High-technology products, then, are seen as the prerequisite for sustaining competition. The better the products of a company are, the higher its competitive advantage. The aim of managing technology, therefore, is to increase the ability of a company to introduce new products more frequently and faster than its competitors do.

Product technology includes several aspects. Effective product planning is necessary to meet customer needs and to define product features and product performance. Without conformance to customer needs, a product is most likely to fail in the marketplace. Additionally, product technology as defined during product development has a strong impact on several aspects, including quality, product performance, producibility, and the plant's potential to customize its products to customer needs. With respect to manufacturing, product development strongly defines the producibility of a product and therefore influences the performance of the manufacturing system. Ease of manufacturing can be achieved through cross-functional integration, reducing the number of different parts, and easier assembly processes. Hence, high performance manufacturers recognize the importance of product development for the manufacturing function. HPM firms consider customer needs as well as the capabilities of manufacturing and of their suppliers in the early stages of product development.

Manufacturing Technology

Manufacturing technology is defined as the equipment and the processes used to make the product. Often the presence of proprietary technology can provide a competitive advantage. The importance of proprietary technology, either patented or nonpatented, should not be underestimated. It is often the only aspect of manufacturing technology attributed to HPM. Nevertheless, there are many other aspects of manufacturing technology that also must be considered, including anticipation of technologies that might be developed in the future. Anticipation could give the plant a competitive advantage by virtue of giving it a jump on competitors.

There are also "soft" aspects of manufacturing technology that could make a great difference in attaining HPM. These include working with suppliers to develop new technology and cross-functional cooperation within the company. When suppliers are included early in the development process, they can suggest new ways to make components or parts. Their suggestions might alter the product design or the design of manufacturing processes. A high degree of internal cooperation, through cross-functional design, is also necessary to achieve an advantage in manufacturing processes or equipment. Cross-functional cooperation helps to clearly define manufacturing requirements during equipment or process development.

Information Technology

The role of information is gaining importance for effective manufacturing management. Information technology is the basis for many concepts directly related to manufacturing at the plant level. Computer integrated manufacturing (CIM), with its components computer-aided engineering (CAE), computer-aided design (CAD), computer-aided manufacturing (CAM), and computer assisted process planning (CAPP), would not exist without computers and information technology. Furthermore, information technology, in general, and computers, in particular, influence all activities in a plant, not only manufacturing. Communication processes can be accelerated and planning systems can be improved through improved availability of and access to data. Simultaneously engineering the products of a plant and their producibility is unthinkable without comprehensive and effective use of information technology.

Integration of Technology

Traditionally, these three facets of technology—product technology, manufacturing technology, and information technology—have been seen as distinctive and separated, as shown in the left part of Figure 6.1. But this seems to be insufficient. The different aspects of technology are closely interrelated and influence each other, as in the right part of Figure 6.1. For example,

Figure 6.1
Views on Technology

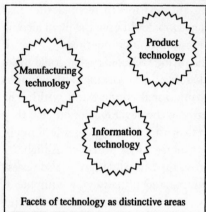

Facets of technology as distinctive areas

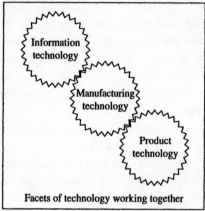

Facets of technology working together

computer integrated manufacturing (CIM), as a part of the manufacturing system, is impossible without information technology; product design and product technology strongly influence producibility in manufacturing and define the manufacturing technology required and CAD is a means to speed up the process of development of new products, leading to improved product designs. As a result, it seems that technology becomes a competitive weapon only if all technology dimensions are linked together in the manufacturing system of a plant. (We do not explicitly consider information technology here but treat it separately in Chapter 7.)

The foregoing discussion shows the strong interrelationships among the dimensions of technology, and it suggests an integrated view of technology. However, even with this integrated view, it is insufficient to be the technological leader in an industry. In HPM, technology is not an end in itself. Products, manufacturing processes, and supporting information technology must "fit" the plant. This means there are people in the plant who use manufacturing and information technology and who design product technology. The employees make the technology in all parts of the plant work. People use information technology to collect information; employees design new products and manufacture them in time, with low costs, reliable quality, and satisfaction of customers needs; and workers and engineers constantly improve products and processes. The organizational structures support technology at all levels of the organization and vice versa—technology supports the processes in the organization itself.

The effectiveness of HPM practices is closely interrelated with technology and influences the success of the technological system of a plant. Together, HPM practices and technology drive performance and competitiveness. Figure 6.2 shows these linkages, which will be discussed later more in detail.

TECHNOLOGY PRACTICES AND TECHNOLOGY IN USE

In order to determine which technology practices lead to improved performance, we need to first define how technology will be measured. We measured technology practices using the scales we developed to assess each of the practices we define here. It is assumed that high performance plants

Figure 6.2
Linkages among Technology, HPM Practices, and Competitiveness

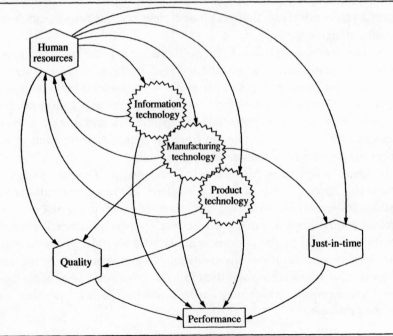

follow these practices more intensively than low performers do, but this assumption will be checked after the definitions are given.

Product Technology

• *Producibility Efforts:* Product development and design determine to a large extent the producibility of a product. They define which material is used, which machines are necessary to process the material, and how easily the different parts of a product can be assembled.

• *Product Design Simplicity:* Closely related to the producibility efforts in the design phase is the simplicity of the product design. A low number of active parts in a product indicates a simpler product design, which allows better control of the manufacturing processes, which leads to lower work-in-progress inventories and easier manufacturing.

• *Phase Overlapping:* With the increasing speed of new product introduction, it is becoming more and more important to accelerate the new product development processes. A plant's ability to introduce products faster than competitors do can be crucial for its survivability. Simultaneous engineering is commonly seen as a means for shortening the product development time of new products. The scale "Phase Overlapping" measures a plants efforts to shorten product development time through overlapping design stages.

• *Interfunctional Design Efforts:* To ensure producibility and product simplicity, manufacturing must be involved in the new product development process in early stages. This requires cross-functional communication and cooperation. Manufacturing has to carry and communicate its knowledge about process potential and capabilities across functional boundaries and influence product design from the manufacturing point of view, from the beginning of the development process.

• *Quality Efforts in New Product Development:* Product quality is an important success factor in the marketplace. However, potential product quality is determined, to a large extent, in the design phase of a new product for several reasons. First, product development influences the product quality perceived by the customers directly through design specifications and quality considerations. In addition, product development influences the ease of manufacturing and therefore the proneness to faults during the production process, which may not be discovered before a product is sold to the customer.

The management practices just described especially concern product development. However, there are management practices closely related to the manufacturing process technology, as well

Manufacturing Technology

• *Willingness to Introduce New Technology:* The effective use of more advanced manufacturing technology and the production of innovative products can be thwarted by the employees if there is a strong resistance toward new technologies. Technology implementation often means rationalization and, hence, an increasing fear of layoffs, which causes resistance. High performance manufacturers show a higher willingness to adopt process and product changes and gain competitive technological advantages from the strong support of the organization.

• *Anticipation of New Technologies:* As new technologies become available, it is thought that plants that anticipate their availability are better prepared to implement them and to use them as a source of competitive advantage. Anticipation of new technologies determines whether the plant is prepared, in advance of technological breakthroughs, to engage in the implementation of new technologies when they become available.

• *Proprietary Equipment:* Proprietary equipment provides a means to gain and to sustain a competitive advantage. Proprietary equipment gives a plant the chance to produce more efficiently and to keep this advantage over a longer period of time because it uses manufacturing equipment that is not available to the competitors. In addition, companies that develop proprietary equipment know more about it than even equipment suppliers.

• *Working with Technology Suppliers:* If the resources of a plant are insufficient to develop proprietary equipment, close cooperation with technology suppliers can be a source of new and more efficient manufacturing equipment. This scale is designed to assess whether the plant works closely with its suppliers of equipment in developing new and appropriate process technology. However, in many cases, the competitive advantage resulting from cooperation with technology suppliers is less effective than the practice of proprietary equipment. This is because technology suppliers may—at least after a certain period of time—try to sell their knowledge to other plants.

• *Effective Process Implementation:* Having installed new manufacturing technology is not sufficient; the new technology has to be implemented appropriately. This should be a critical factor for the effectiveness of manufacturing because the implementation of new manufacturing technology can change the production processes used, may require more skilled employees, and influences the relationships and flows between separate areas in the manufacturing function.

These technology practices should show a strong impact on the competitiveness of manufacturing and product technology and, hence, should lead to a competitive advantage. How these practices work together and how they drive perceived competitiveness will be shown, after we discuss how technology practices vary across national boundaries.

COMPARING TECHNOLOGY APPROACHES ACROSS COUNTRIES

Product Technology

Using the practices that were defined in the preceding section, we compare the product technology approaches used across the five countries in our HPM database. First, we note there are indeed differences in how the plants in each country utilize product technology practices. For example, in designing products for producibility, the plants in Japan have a higher level of practice than does any other country (see Table 6.1). This is not unexpected, for Japanese plants are known for the attention they pay to manufacturing during design. For example, in Japan engineers are frequently rotated between engineering and manufacturing functions. This leads to products that are more easily produced from the initial designs that are provided.

The Japanese plants also are leaders in product design simplicity. In this case, however, the German and Italian plants are also quite good, followed by plants in the United States and the United Kingdom. One example that comes to mind is the approach used by the Japanese for automobile design. Japanese cars often have fewer options and accessories than U.S. and European cars. As a result, Japanese automobile designs are less complex than their competitors.

In interfunctional design efforts, the Japanese, German, and U.S. plants are all quite high, but the Italian and U.K. plants are behind. We believe that the Italian and the U.K. plants have not been as aggressive in applying interfunctional practices to their design efforts, perhaps because they take other approaches instead. We have roughly the same comparison,

Table 6.1
Comparison of Technology Practices across Countries*

	Germany	Italy	Japan	United Kingdom	United States	Significance
Product Technology						
Producibility efforts	3.41	3.40	3.86	3.34	3.40	**0.000**
Product design simplicity	3.42	3.37	3.75	3.14	3.17	**0.000**
Phase overlapping	3.43	3.06	3.59	3.55	3.50	**0.000**
Interfunctional design efforts	3.36	3.13	3.48	3.21	3.40	**0.008**
Quality efforts in new product development	3.80	3.72	3.88	3.86	3.64	0.171
Manufacturing Technology						
Willingness to introduce new technology	4.32	2.49	2.22	2.88	2.16	**0.000**
Anticipation of new technologies	3.74	3.41	3.60	3.41	3.49	0.114
Proprietary equipment	3.04	2.99	3.25	2.74	3.12	**0.043**
Working with technology suppliers	3.38	3.27	3.40	3.48	3.64	0.085
Effective process implementation	3.66	3.44	3.76	3.46	3.45	**0.004**

* The numbers in the table are country means based on scale scores, with 5 being the highest level of practice and 1 being the lowest. The Sig (Significance) of the differences between countries is shown in the last column. A difference of about 0.35 in the means is needed to detect a significant difference between any two countries.

with only the Italians behind, when we consider phase overlapping in the design process.

Finally, the practice of emphasizing quality efforts in new product designs is one that is stressed in all five countries. There is no particular difference among countries, and they all score high on this dimension. We believe this occurs because manufacturers everywhere realize the extreme importance of designing quality into their products from the start. Quality has become a design requirement they all now meet.

In summary, the Japanese plants are ahead of, or at least even with, the other countries in most practices related to new product design technology. This does not come as a surprise; the Japanese have been stressing product design in recent years, while they also continue to pursue integrated interfaces between design and manufacturing.

Manufacturing Technology

In the manufacturing technology area, we see quite a different story. The German plants are leading in many areas, however, Japanese plants are also still very strong. For example, in the willingness to introduce new technology, the German plants have a very strong lead over all the other four countries (see Table 6.1). German plants have a long history of stressing technology in their manufacturing operations, and they reinforce this through their education and training systems. Germans are well known for their fascination with equipment and technology that is actually implemented in their plants.

In the area of anticipation of new technologies, German plants again lead the way, but Japanese plants are a close second and the other countries also do quite well. So, German plants are more willing than other countries to introduce a new technology, but they do not anticipate it to a much greater extent than in the other countries. As a matter of fact, no country can claim to be the best at anticipating new technology, in the rapid age of technological change in which we live today.

Aside from a willingness to introduce new technology and anticipation of new technology, there are some differences in what is actually implemented in the various countries. For example, the Japanese plants emphasize proprietary equipment to a somewhat greater degree than do the plants in other countries, and particularly those in the United Kingdom. The emphasis on proprietary equipment can lead to a competitive advantage in both products and processes.

On the other hand, the U.S. plants lead in the area of working with their technology suppliers. This can also lead to an advantage, particularly if the plant has exclusive rights or secrecy agreements with its technology suppliers.

Finally, we find that effective process implementation is one area in which plants in both Japan and Germany lead those of the other countries. It is important not only that plants anticipate and plan for new technology, but also that they are able to effectively implement it. This is, therefore, an advantage for the Germans because they are not only more willing to implement new technology but also are good at effectively implementing it in their plants.

In summary, we note that both the Japanese and the German plants are stressing technology management. The Japanese plants have the lead in some of the new product technology areas, whereas German plants are leading in manufacturing technologies. Nevertheless, each plant, regardless of the country, should adopt a technology strategy to fit its own situation and circumstances. Simply being the leader in new technology is not always the best practice for a particular plant. There are tendencies, in certain countries, brought about by their past practices, culture, educational systems, and other factors that create differences among countries, as well as differences among plants.

TECHNOLOGY PRACTICES AND COMPETITIVENESS

The practices described in the last section are used to investigate the impact of technology on competitiveness. In addition the perceived competitiveness was measured for both product technology and manufacturing process technology. This distinction is important because different technology practices may show a distinct impact on product technology performance, on manufacturing process technology performance, or on both.

Competitiveness of Product Technology

As shown in Figure 6.3, a plant's willingness to introduce new technologies shows the strongest influence on the competitiveness of product technology. Effective implementation of new product technology also shows a significant influence on the competitiveness of product technology. The smoother the production processes—due to effective use of manufacturing technology—the more time there is for the improvement of product technology. Additionally, the design simplicity of a plant's products is significantly related to the competitiveness of product technology. Surprisingly, the practices of interfunctional design efforts, the overlapping of design phases, and the quality efforts in new product development processes show no significant influence. This could occur because these practices are interrelated to other practices that were found to be significant. Because of this, these practices explain only 35 percent of the variance of the competitiveness of product technology.

Competitiveness of Process Technology

Regarding the competitiveness of manufacturing process technology, effectively implementing new process equipment, intensive efforts in producibility during the design stage, and anticipation of new technologies have the strongest impact. The relationship between proprietary equipment, product design simplicity, and the competitiveness of manufacturing technology is not as significant as the other practices. Because both practices also explain some of the variance of the competitiveness of process technology, they are

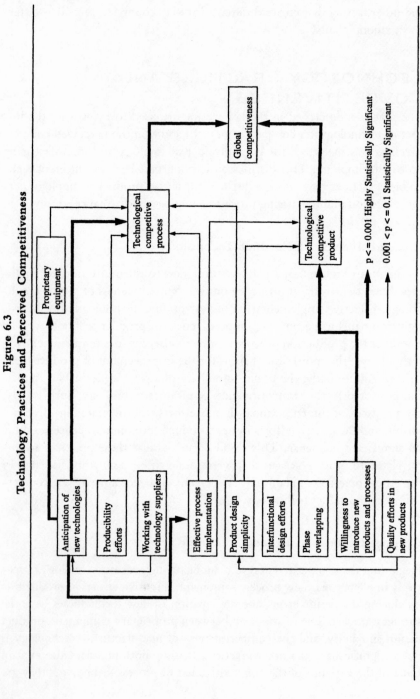

Figure 6.3
Technology Practices and Perceived Competitiveness

104

still considered in the framework. Again, interfunctional design efforts and phase overlapping show no significant influence.

What this analysis clearly indicates is that willingness to introduce technology, anticipation of new technology, and effective process implementation are the keys to competitive performance. Proprietary equipment or practices, such as interfunctional design or design for producibility, are not by themselves as significant. This shows that a general climate for new technology, followed by aggressive implementation, is the key, rather than the use of any specific practice.

One plant we visited showed how this can work. It continuously scanned the environment for new technology. Its customers relied on the plant to design and manufacture leading-edge products. That plant's anticipation of technology and willingness to introduce new technology clearly gave it an edge over its competitors. Of course, it also patented its products and processes and used cross-functional design teams, but that did not give the plant an edge. Its key to success was its image in the marketplace as a leader that was willing to introduce new technology, followed by aggressive implementation of new processes.

DOES TECHNOLOGY LEAD TO BETTER PERFORMANCE?

Figure 6.4 shows a technological competitiveness matrix, with the dimensions of perceived competitiveness of product technology and competitiveness of manufacturing process technology as a basis for a classification of the plants. The technological leaders with highly competitive product and process technology are located in quadrant 1. Quadrant 2 shows plants focusing on high competitiveness of process technology, and quadrant 3 shows the plants with a focus on manufacturing product technology. The plants with below average competitiveness of product and process technology, described as technological underperformers, are in quadrant 4. Classifying the plants based on this matrix is a good indicator of membership in the high- or low-performing cluster of performance. As Figure 6.4 shows, the companies with high performance overall are mostly located in quadrant 1 of the matrix. In this quadrant, the plants perceive the competitiveness of their product technology and manufacturing process technology as being above average. The plants with low overall performance are mainly located in the technological underperformer quadrant (quadrant 4), as well as in quadrants 2 and 3, where plants focus on a single dimension of technology.

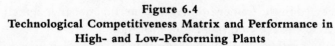

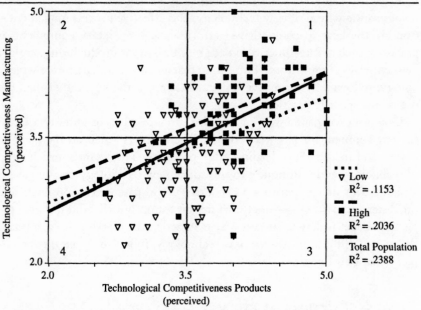

Figure 6.4
Technological Competitiveness Matrix and Performance in
High- and Low-Performing Plants

Although classification in the matrix does not perfectly separate high- and low-performance manufacturers, technology clearly plays an important role in achieving a position as a high-performing plant. Since other practices, like human resource management, strategy, or quality, also influence a plant's effectiveness, it is not surprising that a plant's position in the technology matrix does not perfectly explain the cluster membership.

Additionally, plants that concentrate on competitiveness in product technology while neglecting process technology competitiveness are more likely to be low-performing plants. Also, plants that focus on competitiveness in manufacturing technology and disregard their product technology competitiveness are more likely to be low-performing plants. Therefore, it is best to simultaneously concentrate on both dimensions of the technology competitiveness matrix.

TECHNOLOGY IN HIGH-PERFORMING AND LOW-PERFORMING PLANTS

As shown in the preceding section, the use of technology practices has an impact on competitiveness. But are there any specific differences that clearly distinguish high-performing and low-performing plants?

Considering the different technologies that are in use, we found that high performers are more innovative than low performers and are more likely to introduce innovations sooner than the low performers. As shown in Figure 6.5, the high performers introduced computer-aided design, computer and direct numerical-controlled (CNC/DNC) manufacturing equipment, and flexible manufacturing systems (FMS) about 18 months earlier than the low performers did.

The high performing plants also had newer equipment. Only 40 percent of the plant equipment is newer than 5 years in the low-performing plants, whereas in high-performing plants, 50 percent of the manufacturing equipment was purchased within the past 5 years. Comparing the percentage of

Figure 6.5
Year of Introduction of Technology for High- and Low-Performing Plants

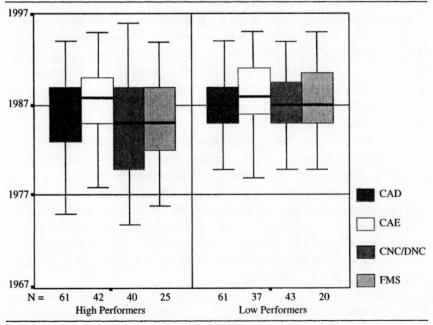

equipment that is newer than 2 years, it is obvious that high performers are investing in new machinery more intensively. Almost one-fifth of the equipment was purchased in the past 2 years in high-performing plants; in low-performing plants, however, only one-tenth is newer than 2 years.

A good example for the use of new and proprietary equipment is PVT Siebler, Remchingen, a German manufacturer of shrink-wrapping packaging machines. It built machines for automated packaging of high-pressure and temperature-sensitive goods, like pills. PVT Siebler has a clear competitive advantage in the manufacturing of critical components of its packaging machines. Although it uses outside engineering capacity to a large extent, the manufacturing equipment for the critical components is completely designed and built in house, and equipment is kept up-to-date. For Siebler's management, this is one of the reasons it is able to maintain its outstanding competitive advantage in product technology and to defend a market share of 70 percent.

Our analysis suggests the importance of technology-related management practices. The effective implementation of new processes, considering producibility in the design phase of new product development, product design simplicity, and a plant's ability to anticipate new technologies were found to be important practices, improving the competitiveness of process technology.

However, these technology-related practices will not work without a proper setting in the plant. Clearly defined and strong strategic orientation, employee commitment, a climate that is open to improvement suggestions, task-related training of employees, and their multifunctionality and their breadth of experience are all important catalysts to make technology work and to let the different functions in a plant play together as an ensemble. Balancing the efforts is more important than being strong in most of them while failing in some strongly supporting practices.

The idea of balancing is also supported when the measures of the newness of equipment and the percentage of sales from products introduced in the past 5 years are plotted against each other in Figure 6.6. In principle, low product innovation goes along with low innovation in manufacturing equipment. As far as performance is concerned, we conclude there is a stronger tendency for high performers to have a balanced relationship between product and process innovation. Product and process technology play well together.

High-performing plants use a significantly higher percentage of new manufacturing equipment. This could lead to the conclusion that investment in

Figure 6.6
Balancing Product and Process Technology

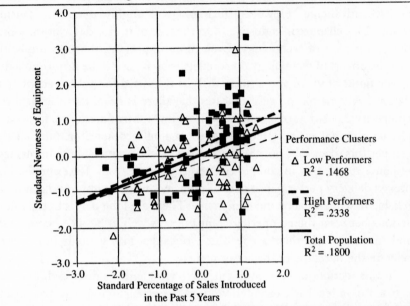

new, more sophisticated manufacturing equipment automatically will improve a plant's competitive position. New equipment improves several performance measures, but a high percentage of new manufacturing equipment has to be actively maintained by a plant. As time goes by, equipment ages and, hence, the newness of the equipment decreases. For example, one of the high-performing plants in our sample indicated that 50 percent of its equipment was newer than 2 years. In order to keep such a high percentage of new equipment, an average growth rate of sales of about 60 percent is necessary, assuming that 10 percent of the sales is used to invest in manufacturing technology. However this particular plant only invested 2.9 percent of its sales in new equipment, and the growth rate of the market was only 4 percent. Hence, the newness of its equipment is very likely to decline. To use such a high-performing plant as a role model to identify successful management practices could be very dangerous. The broad view, based on our analysis of plants, and a dynamic view are necessary to derive recommendations for management practices that lead to outstanding manufacturing performance.

THE DYNAMIC VIEW OF TECHNOLOGY

The appropriate use of technology is an important means to achieve a competitive advantage. However, technology is highly dynamic, causing tremendous changes in industries. On the face of it, this description shows the importance of technology. Even more important, it shows implicitly that the effects of technology on competitiveness have to be seen from a dynamic point of view, and a static analysis is limited for policy recommendations. A dynamic process indicates that there is no discrete shift in the competitive position of a company through the introduction of, for example, new manufacturing technology by a single competitor. Neither the competitors' decisions nor the markets and the customers show immediate response to changes in competitiveness that result from, for example, improved delivery performance of one company caused by improved and more reliable new manufacturing equipment. There are long delays in the interactions between the competing companies, as well as between companies and customers; and there are feedback loops that may strongly limit or accelerate the outcome of the actions taken.

In this section, a more comprehensive view will be discussed, one that is based on a feedback-oriented view of the problem and that uses terminology coming from the system dynamics approach.* This approach is appropriate for dealing with dynamic and feedback-driven processes. The basis for the conceptualization of the model is the information and data collected within the HPM project, as well as logical analysis. The starting point of the feedback structure, shown in Figure 6.7, is the performance of manufacturing technology as defined in the previous section. The feedback reinforcing loop $R1$ causes a process of either exponential growth or decay. Increasing manufacturing technology performance should improve flexibility, delivery, and the percentage of products delivered to the customer without defects. These measures increase customer satisfaction and cause the market share, as well as sales volume, to rise.

Since sales volume is increasing, the investment budgets and investments in manufacturing technology must increase as well. This causes the newness of equipment to improve, and finally the manufacturing technology performance grows because newer equipment is more sophisticated than older equipment. However, these "investments in manufacturing technology" loops have time delays, because investment decisions have to be made

---

* For more details, see Forrester (1961, 1971) and Sterman (2000).

Figure 6.7
Feedback Loops and Limiting Driving Technological Competitiveness

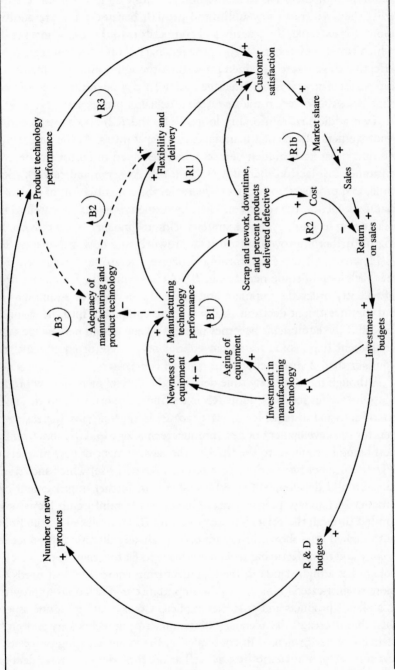

and new equipment has to be ordered and built up in the plant. Undoubtedly, they will not show unlimited growth because there are additional loops, for example, the reactions of competitors to the loss of market share, which is not shown here. Hence, the feedback loop R1 is not seen as being effective to generate sustained growth. Although it improves manufacturing performance, it is not effective in day-to-day operations because of the time delays involved in the investment decision processes.

Two additional reinforcing loops (R1b and R2) also cause continuous improvement of a plant's manufacturing capabilities. As the percentage of products that have hidden defects when delivered to customers decreases, customer satisfaction increases (loop R1b). As scrap and rework, downtime, and products with defects decrease, the cost situation of a plant will improve. Hence, the return on sales, investment budgets allocated to manufacturing technology, and manufacturing technology will improve, leading to further improvements in scrap, rework, and downtime (loop R2).

However, all three reinforcing processes are strongly balanced by the feedback loop of equipment aging, B1. As investment increases, the amount of recently ordered equipment and hence the newness of equipment rises; but the investment decision also puts more equipment into the aging process, and the equipment becomes older. To keep a high percentage of new equipment (e.g., some plants have about 50 percent of equipment that is younger than 2 years), enormous growth is necessary.

Although there are long time delays in the development of new products, the reinforcing feedback loop R3 is the most important loop to generate continuous and sustainable growth processes in the long run. Increasing budgets for the development of new-product technology lead to more new products being introduced to the market, the development of new markets, and higher customer satisfaction. As a result, sales will rise, which increases the research-and-development (R&D) budgets for further improvement of the product technology performance. However, the reinforcing loops will be limited through the balancing loops B2 and B3, introduced by the linkages with dashed lines shown in Figure 6.7. As already discussed, product technology and manufacturing technology have to fit together. If product technology has a high standard, the manufacturing equipment also needs to be more sophisticated. This makes the importance of a balanced improvement in a plant's products and processes explicit. Concentrating on one aspect or the other decreases the adequacy of manufacturing technology and product technology performance. In consequence, the manufacturing processes run less smoothly, and flexibility, as well as on-time delivery rate, decreases.

Delivery time, scrap and rework, and downtime increase, and more products with hidden defects will be delivered to the customers. These balancing loops are able to undermine, to some extent, the effects of the reinforcing loops. The inadequacy of product and manufacturing technology can get worse if the positive feedback loop R3 is dominating the system over a period of time. Because improved product technology leads to higher customer satisfaction, market share, sales, and return on sales and therefore is strengthening the investments in product technology, further improvements of product technology performance cause the fit of manufacturing and product technology to decrease.

Furthermore, the technological system of a plant does not exist in a vacuum. The technological system has to be embedded in and interact with the human resource system. Products, manufacturing processes, and supporting information technology, as well as the performance of human resources, have to fit together. Without investment in skilled employees, growth processes initiated by higher investments in manufacturing technology or product technology are likely to fail from the very beginning. The missing link between technological strategies and other areas of a plant, in particular human resources, is an important cause of failure.

In general, it was found that investment in manufacturing technology enables a plant to gain some market share. However, continuous long-term growth cannot be achieved by a strategy that focuses only on manufacturing technology—unless the market grows by external forces. Sustainable, internally influenced growth results only from new products technology and the development of new markets. Therefore, it is recommended that manufacturing technology improvements should be used to support the growth processes stemming from new product technology.

CONCLUSIONS

When setting a technology strategy, it is important to consider not only product technology, but also manufacturing technology and information technology. Only by integrating these three types of technology can HPM be achieved.

Willingness to introduce new technology, anticipation of technology, and effective process implementation are the keys to HPM. While buying new equipment is also important, the approach used to introduce technology will be critical. It is important to invest in both manufacturing technology and new-product-introduction technology. Either one of these investments by itself is less effective than both approaches used together. This is because new products and manufacturing technology tend to reinforce each other and have a synergistic effect.

Different countries are using different approaches to technology development. Plants in Japan are stressing new product introduction including producibility, product design simplicity, and interfunctional design. These practices are streamlining and integrating the new-product-introduction process. In contrast, the German plants are excelling at manufacturing technology, including willingness to introduce new technology and effective process implementation. Simply being the leader in one area of technology or another is not enough; each plant must develop its own technology strategy to compete in the future.

The technology strategy defines the path that the plant should follow. It deals with such issues as the appropriate mix of new product and new manufacturing technology, whether the plant should lead or follow the competition, and the approach used to implement new technology. A clearly defined and communicated technology strategy is the key to competitive advantage.

REFERENCES

Forrester, Jay W. *Industrial Dynamics*. Portland, Oregon: Productivity Press, 1961.

———. *Principles of Systems*. Portland, Oregon: reprinted 1990 by: Productivity Press, 1971.

Porter, Michael E. *Competitive Advantage—Creating and Sustaining Superior Performance*. New York: Free Press, 1985.

Solow, Robert M. "Technical Change and the Aggregate Production Function," *The Review of Economics and Statistics* 39 (1957): 312–320.

Sterman, John D. *Business Dynamics: Systems Thinking and Modeling for a Complex World*. New York: Irwin/McGraw-Hill, 2000.

CHAPTER 7

INFORMATION TECHNOLOGIES FOR HIGH-PERFORMING PROCESSES

CIPRIANO FORZA, KATHRIN TUERK, and OSAM SATO

$\mathbf{T}$ony Salvatori shook his head in frustration. He had just finished reading a trade journal on advanced computer applications for manufacturing. Although he had suspected that his plant was behind its competitors in information technology, he had no idea just how far behind it was. His German and U.S. competitors, in particular, were in the process of installing very sophisticated (and extremely expensive) manufacturing information systems, such as Enterprise Planning Systems (ERP), that linked financial, marketing, human resources, and other databases in real time. He knew his company would never be able to afford an investment of that magnitude, although the company had a substantial amount of funds available for investment in information systems. How would he ever be able to compete globally without a state-of-the-art computer and information system?

THE CHANGING ROLE OF INFORMATION SYSTEMS IN HIGH PERFORMANCE MANUFACTURING

The role of information systems for manufacturing has undergone dramatic changes during the past few decades. The early use of information technology (IT) concentrated on the automation of individual activities, with the aim of increasing their efficiency and reducing their costs. Automation was therefore aimed at improving the execution of individual operations, while the contribution it made to coordination was decidedly marginal.

During the 1970s, innovations in the IT field began to make it possible to go beyond the previous level of isolated automation, leading to the birth of computer-integrated manufacturing (CIM). With CIM, information technologies were no longer viewed as solely powerful instruments for the automation of single operations, but they were seen as the bearers of an inter- and intraplant integration model, based on the automation of all aspects of information exchange.

During the 1980s, the predicted scenario was for a future populated with unmanned factories, which were to be highly automated and integrated both internally and externally; however, this has not been realized. Implementation of this model was punctuated by a series of failures and of results that fell below expectations. In addition, in HPM, the meaning of the term *integration* is radically changed: it is no longer an *a posteriori* way to mend the fractures created by the old structures and technologies, rather it is redesigned *a priori* as a lever to dissolve the barriers between the various activities involved in a particular process.

Schonberger, one of the first authors to present this approach in the West, emphasizes the reduction of information system complexity due to flow production; reduction in the number of parts; and the use of just-in-time practices (JIT), quality management, and employee involvement. The Schonberger vision of simple information systems and fewer computers in manufacturing derives from his attention to simplicity. However, he also recognized the increasing importance of IT in design/engineering and in building up a paperless chain of customers.

Although the vision of the CIM model started to decline during the early 1990s, both the potential and the adequacy of IT increased dramatically during that period. Some studies began to foresee a new role for IT: An MIT research project highlighted that IT is making possible fundamental changes in the way production, coordination, and management work are done. We believe that IT is only an enabler, however; to actually change the way these activities are carried out takes a combination of managerial leadership and employee participation.

Successful application of IT requires changes in management and organizational structure. IT is a critical enabler of the re-creation/redefinition of the organization because it permits the distribution of power, functions, and control to wherever they are most effective. Therefore, the advanced use of IT requires continuous improvement and is characterized by leadership, vision, and the sustained process of organizational empowerment. This idea, which emerged from MIT studies, has become more widespread as business process reengineering (BPR), with its focus on processes, has become popular. In 1993, Davenport proposed a radical change of outlook, stating that mere investment in information technologies cannot produce a return in terms of performance. It is the resulting changes made to the processes that can produce these benefits. Thus, it is being recognized that use of information technologies and new organizational models are complementary.

High performance manufacturing is the result of the linkage between three innovation clusters: (1) process automation technologies, (2) information integration technologies, and (3) new organizational management techniques. In this renewed meeting between the IT and production models, the role of information technologies is changed from the CIM perspective of the 1980s. The huge variability and uncertainty of the competitive environment in which plants have to operate today, combined with the flexibility that they need, are not compatible with the static vision proposed by traditional CIM models, which suggests the achievement of improved performance by modifying the automation of activities and information exchanges while leaving process characteristics unchanged.

In HPM, the emphasis is transferred from the integration of IT applications, systems, and machines to plant processes. These processes are redesigned and integrated into the new organizational and management systems, with the help of IT. Information technologies then acquire a support role, functioning as one of the levers through which improvement can be achieved. The CIM approach is replaced by a new vision, which aims at developing an information system that can be easily adapted to future information needs, which may not even be identifiable at the moment. Thus, both the computerized component and the noncomputerized component of the information system should be emphasized in order to offer support to the realization of high-performing processes (see Figure 7.1).

The remainder of this chapter elaborates on the implications that HPM practices have on information systems, in relation to the main operating processes: product/process development, physical transformation processes, and logistic processes. Part of the administrative process is also considered, since target setting and performance measurement have been highly affected by new managerial practices and are highly interconnected with operations.

INFORMATION SYSTEMS FOR NEW PRODUCT/PROCESS DEVELOPMENT

Product/process development has a special significance in HPM. Not only is product/process development the fundamental linkage point between customer needs and the realization of products that satisfy those needs, but

Figure 7.1
IS Support to High Performance Manufacturing

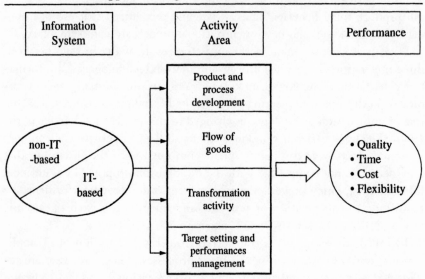

it is also the main means of determining time to market and the complexity of production activities. In order to develop the process capability for achieving high quality, reduced time to market and lower levels of logistic and production complexity, high-performing plants have modified the product/process development process, including the introduction of information technologies and changes in information flow.

Information Systems for Design Quality

Satisfying the customer means proposing products and services that live up to or exceed their expectations. This requires the fast identification of customer expectations and translating those expectations into product and service specifications that conform to or improve on expected quality. To ensure that the identification of customer needs is reliable, various communication channels with the customers should be organized, such as free telephone numbers, Web contacts, visits, and questionnaires. To facilitate translation into project specifications, every effort should be made to ensure that the information does not arrive at the designer in a distorted form. Direct meetings with the various people involved and information systemization tools, such as the house of quality and electronic data exchange about the project, are used in order to ensure the nondistortion of design information.

The case of a plant that we visited that designs electronic control devices illustrates this. A particularly able project leader appoints designers who will be involved in the design process to follow the meetings on the new product right from the very first meeting. In cases where the customer's ideas are not sufficiently delineated, the project leader asks the designers to speak directly to the customer. The ability to see where the electronic control device will be physically inserted and the environment in which it will be used greatly helps designers to interpret the provided specifications in a meaningful way. Communication with the customers during the development phase not only facilitates the understanding customers' needs, but also, in some cases, makes it possible to *guide customers' expectations*. These visits are used therefore as a means for reducing the distance between perceived quality and design quality.

A well-designed information chain has a strong positive impact on the design process, in allowing design quality to more closely conform to expected quality. We found that the plants with high performance levels had a higher level of information exchange with customers, both explicitly during new-product development phases and in general, by continually obtaining feedback on quality. This leads to greater conformity and, therefore, to greater customer satisfaction.

Information Systems for Faster Design

The HPM development process aims at being structurally faster. This makes it possible to postpone the starting point of the development process in order to collect greater quantities of customer information, to carry out greater experimentation so there are fewer problems during the production process, or to arrive first on the market, thus obtaining a premium price or an advantage in terms of market share. In order to do this, high-performance manufacturers take multiple actions, such as speeding up single activities and modifying connections between the various activities involved, including product conception, definition, design, detailed production, preparation, and realization and postlaunch design modification.

Acceleration of various activities and information flows between the phases not only reduces the total development time, but also lowers the costs of engineering changes so that project modifications become less costly. Computer-aided design (CAD), computer-aided engineering (CAE), and computer-aided process planning (CAPP) give operators new abilities in terms of the speed with which these activities can be carried out. The use of such tools is not sufficient, however, to guarantee higher performance levels, as can be seen in Figure 7.2, where we found few differences between high- and low-performing plants, in terms of their use of CAD, CAE, and CAPP. Better performance derives from the *appropriate use* of the various instruments and the ensuing *modification of the design processes.*

In HPM, there is a tendency to overlap design/engineering activities to cut time to market and to improve the quality of design. This is accompanied by bidirectional communication to avoid early starts in the dark; however, information on the design specifications cannot be very detailed at this point. In order to avoid heavy constraints on subsequent phases or changes to decisions already made, HPM design relies on the involvement of all functions from the beginning of the project. This includes face-to-face meetings not only between designers who are involved in various phases, but also between people with different functions that have a fundamental role during the life of the project.

In HPM, product/process development is not limited to the plant. Information from customers and from suppliers of components and technology is actively sought and used. Collaborative relationships are formed, even to the point of co-design of some components. In this way, information about modifications in design characteristics arrives in time to avoid

Figure 7.2
IT Use in Design/Engineering

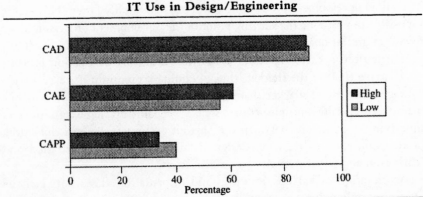

Table 7.1
Horizontal Information Flow for Design in High-Performance Plants

	p-value	High Performance	Low Performance
Quality information from suppliers	0.03	3.88	3.67
Quality information from customers	0.00	3.87	3.62
Collaboration with technology suppliers	0.00	3.50	3.25
Multifunctional team in design	0.00	3.50	3.19

costly design changes. Information on product/component problems, production constraints, innovation possibilities, and so forth, therefore circulates not only within the plant, but also between the plant and its supply chain. Table 7.1 shows that we found important differences between high- and low-performing plants in the amount of horizontal information.*

The design phase in HPM plants is therefore characterized by both greater *internal integration* (between the design department and the other functions and among the various activities in the design department) and *external integration* with customers and suppliers. This high level of interaction should be supported by an information system that is oriented toward networking. We found a difference between high- and low-performing plants in the presence of local-area networks (LANs) in design: 74 percent of the high-performing plants used LANs in design, in comparison with 63 percent of low-performing plants. However, a higher difference was found in the use of electronic connections for the exchange of drafts, drawing, and design data (Figure 7.3). Thus, the availability of LANs in the design department and of electronic connections with other areas of the plant and with customers and suppliers for the exchange of design data, designs, graphics, and so on, speeds up information exchanges and makes them more reliable.

We also found that although high-performing plants tended to introduce both simultaneous engineering and information technology to support design, most low-performing plants tended to computerize design without introducing simultaneous engineering. This provides further evidence that in HPM plants, information technologies are used as support for innovative managerial methodologies, not simply as instruments of automation. Thus, HPM plants make the decision to install new technology with an explicit orientation toward a redefinition of the design/engineering process.

---

* Differences that have a p-value of 0.05 or less are considered statistically significant differences.

Figure 7.3
Percentage of Design Data Exchanged Electronically

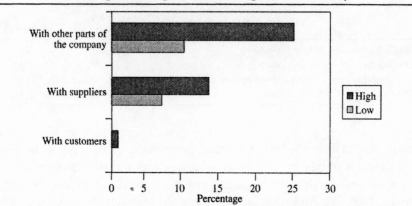

Design Information to Reduce System Complexity

HPM requires a design process that is oriented toward simplicity. This can be pursued through commitment to the development of projects that make production and assembly easy by requiring a low number of parts, preferably the same ones for different products, making assembly based on customer orders possible. The basic principle is reduction of the expensive fraction of complexity, which has less added value from the customer point of view.

Thus, the goals of the HPM design process are (1) *standardization,* (2) *modularization,* and (3) *creation of a platform.* These three actions have an impact on product complexity and on information systems in the design and production/logistics processes. On the one hand, these approaches require more highly evolved design information systems to help designers follow these practices. On the other hand, by increasing the communality between the various products, new design methods become possible, supported by information systems that exploit these common qualities. To support standardization, for example, it is first necessary to create a file for the storing of information on parts and components. When a new project requires a certain component, the archive for that component can be consulted to see if an appropriate version already exists.

Design activities supported by such information structuring and storing can use computer support much more efficaciously. For example, having the possibility of associating CAD with a database containing all the models elaborated to that point means that these models can be retrieved and modified, and congruence between existing and needed parts can be verified in real time. Similarly, CAPP, which integrates product and process design, is supported by databases containing product family codes and standard production cycles. After having recognized that a new product belongs to a family, the production process associated with that family can be retrieved and modified, if necessary.

Plants that aim for product customization have to maintain high variety levels, but at the same time they must be fast and also keep their costs down. This is why it is critical for these plants to capitalize on existing knowledge of components and product architecture during the design stage. It is very probable that the designers will have to tackle some problems similar to those already solved in the past. If the solutions identified are left solely to the designers' memories, the time to retrieve those solutions is certainly higher than the time it takes to retrieve information from an information file such as a "lessons learned" book.

The case of several Italian plants oriented toward product development illustrates this. They have developed information systems that make it possible to program new products quickly by starting with knowledge about previous development projects. If customers require customization, the plants are able to easily estimate both the feasibility of the request and the additional costs that the modification would create. In these plants, a high level of customization is reconciled with the need to keep costs down by means of common architectures, allowing the modification of some elements while keeping other elements unchanged. The need to manage these modifications without errors and with high standards of quality requires information systems that easily follow the evolution of the requirements and that can organize the modification in all phases—request reception, design, and production management. The information system is therefore coherent with the focus of these plants on customization, flexibility, and quality.

INFORMATION SYSTEMS FOR THE TRANSFORMATION PROCESS

The transformation process in high performance manufacturers is characterized by a high level of reliability and a low level of variance. It produces components and products not only within the specification limits but also very close to the target values. This capability requires information tools for process control and improvement, an information environment in line with employee empowerment/involvement, and information exchange with customers and suppliers to evaluate and improve the transformation process all along the supply chain.

Information Systems for Control and Improvement

According to the HPM perspective, not only must quality be built into the transformation process, but the ability to achieve quality must also be continually improved. Therefore, reliable transformation processes need the continual application of control and improvement systems.

Process control and improvement is based on a set of information instruments that signal the presence of anomalies and that help to identify the causes of problems. Statistical process control (SPC), which is part of the control-action-control cycle, is particularly useful (Table 7.2). We found that high-performing plants use internal quality information collected through SPC, both in real time and for subsequent analyses aimed specifically at process improvement. It is therefore very important that this information is collected, stored, and made easily available for those who can carry out changes in order to improve process performance.

Information technology support is not strictly necessary for SPC, but it can be useful in a production context, such as that of HPM, where information systems are built into machines. The use of machines that automatically carry out SPC during production makes the use of software for statistical process control more natural, facilitating the collection of a larger data mass, which can be stored in the database for subsequent analysis. Therefore, in HPM plants, information technologies are more frequently used for process transformation and, in particular, for those applications

Table 7.2
Information Flows for Transformation Reliability

	p-value	High Performance	Low Performance
Statistical process control	0.00	3.57	3.23
Internal quality information	0.02	3.67	3.50

that permit process monitoring and offer information for improvement. We found that in high-performing plants, 10 percent of the machinery is generally fitted with an on-line quality-control device, compared with only 1 percent of low-performing plants. In addition to this, 62 percent of the high-performing plants had databases providing quality information about reliability and process control, as compared with 46 percent of low-performing plants.

Of the information instruments for obtaining and improving quality, especially well known and frequently applied are the "seven tools": (1) trend diagram, (2) histogram, (3) scatter diagram, (4) pareto diagram, (5) control chart, (6) flow chart, and (7) cause/effect diagram. These tools are widely used in both high- and low-performing plants; each of the seven tools is used by 70 to 80 percent of the plants, with exception of the scatter diagram, which is used by 52 percent of low-performing plants and by 64 percent of high-performing plants. Some tools (Figure 7.4) are mainly used manually (for example, cause/effect diagrams), while others are often used with IT support (histograms, trend diagrams, and scatter diagrams).

To achieve reliability in the transformation processes, it is necessary to have well-maintained machinery; and so in HPM, special attention is paid to the management of maintenance activities. Thus, it is important to have

Figure 7.4
Percentage of Plants That Use Computerized Seven Tools

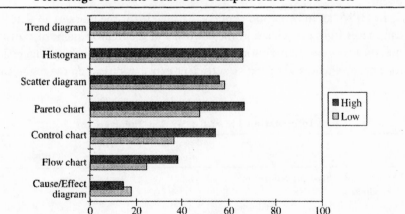

information available on the state of the mechanical resources. Information on which actions have been taken, when the maintenance activities were carried out, and which anomalies had been perceived before the actual breakdown occurred can be collected by the machine operators. Other information can be supplied by built-in diagnostic control devices.

Information Systems for Worker Involvement

In HPM, tasks are enriched by such activities as prevention, monitoring, transformation process evaluation, and feedback. Thus, there is a distribution of responsibilities that modify information needs. In particular, there is a demand for clearer, more accessible information about the ways in which the various activities are carried out and a communication system that encourages employee involvement and empowerment. Table 7.3 highlights some of the differences in information flows for worker involvement that we found between high- and low-performing plants.

High performance manufacturers recognize that workers should be properly informed regarding the activities they must carry out. Hence, procedures in the production departments and instruction manuals should be accurate and easily accessible. Procedures are respected, and they are updated and improved with the help of the workers themselves. We found that in high-performing plants 90 percent of the machines have visible information regarding the state of their maintenance, running efficiency, and fittings, as opposed to 70 percent of the machines in the low-performing plants.

In HPM, an information context is available for the workers so that they can successfully participate in the control and the improvement of transformation processes. Supervisors encourage the workers to supply suggestions for improvement, and they give sufficient explanations when the suggestions

Table 7.3
Information Flows for Worker Involvement

	p-value	High Performance	Low Performance
Use of standardized procedures in production	0.01	3.68	3.39
Suggestions followup and feedback	0.00	3.77	3.46
Problem-solving groups	0.00	3.79	3.53
Quality feedback to workers	0.00	3.55	3.11

are not adopted. Department heads encourage horizontal communication and exchange of opinions and ideas within the problem-solving groups. Vertical communication, which emphasizes the importance of the availability of information for the personnel and of information transparency, can help to establish a climate of trust between the workforce and management. The personnel thus receive feedback on both the quality of the production process and the quality of their work. This type of information environment is useful not only for involving the workers, but also for increasing their knowledge and abilities. This increase in knowledge, pursued through use of SPC, job rotation, and multifunctional training, reduces the need for formal information-exchange activities to be carried out.

Information Systems for the Supply Chain

The quality that the final customer experiences is the synthesis of the quality results for every member of the supply chain. HPM emphasizes the importance of the linkages with customers in order to get feedback on quality levels and on product performance during use so that possible areas for improvement can be identified. In high-performing plants, we found that customers were more likely to visit the plants and to inform them about the quality levels they were experiencing. Customers were also more likely to be included in improvement projects. The result is higher levels of customer satisfaction and reciprocal knowledge.

A symmetrical situation is experienced by the supplier. Any nonconformity should be detected as far upstream as possible, and whenever a defect is found, the point in the supply chain where it originated should be identified. This reduces the presence of defectives in downstream phases and of non-value-added activities associated with defectives generated in the upstream phases. Hence, high performance manufacturers collect information on suppliers so they are selected based on both their quality management and their willingness to improve. They also supply analyses of suppliers' work and provide various types of feedback that permit the suppliers to identify specific areas for improvement. The reverse is also true, with information from suppliers on the results of checks made by them being used by customers as a substitute for their own checks. Thus, information exchange regarding quality of the transformation is not something that stops on the shop floor of each single plant but that extends, in a reciprocal fashion, along the entire supply chain. Information exchange also concerns the suppliers of technology. Information on maintenance work and suggestions for improved equipment operation is important input for the development of machinery that can improve process reliability.

INFORMATION SYSTEMS FOR MANAGEMENT OF THE FLOW OF GOODS

Simpler Physical Flows Allow Simpler Information Systems

The first step to be able to undertake HPM practices in the management of physical flows is the *reduction of complexity*. Complexity, as pointed out by Galbraith, is a function of the number of significant elements that make up the system, the degree of uncertainty (the difference between the information necessary and that available in relation to these elements), and the level of interdependence among the elements. Standardization and modularization reduce the number of elements, and physical contiguity in the location of subsequent activities, greater linearity, and the nonintersection of materials paths reduce the quantity of information that is needed. With initiatives such as plant-within-a-plant and cellular manufacturing, the interdependence among the elements is reduced. The need for information regarding programming and the management of the physical flow is also reduced. Once the system has been simplified, shop-floor-control information systems can also be simplified, for example, through the use of Kanban as an information tool.

Cellular layout is highly effective in simplifying information systems. With cells, a great deal of the bureaucratic work associated with recording the movement of materials can be avoided. Bills of materials are simplified because if several phases are sequenced, it is not necessary to record the intermediate steps. The management of personnel is simplified because the overall results of the cell are rewarded and it is not necessary to get information on the activities of each person. If the cell machines are organized in a U-shape, communication between workers is facilitated through their ability to see what is taking place at the other work stations.

Planning is also simpler because less information is passed to the shop floor. The flow within the cell is self-checked, so planning is done at the level of the cell, rather than at the level of the machine. In setup-time reduction, the levels of the bill of materials can also be reduced because many codes can be treated as ghost codes. Balancing can be done in real time, if the workers are cross-trained.

Short-Term Stability Requires Greater Planning

HPM tries to bring about a timely and stable flow of materials throughout the transformation process. Not only does this reduce stocks, making rapid and punctual deliveries possible, but it also leads to a tighter coupling of the various activities. In order to achieve stability, there must be greater planning. Information systems, if used appropriately, are extremely useful in this regard.

In high-performing plants, long-term production plans are defined to verify overall production capacity and to make agreements with suppliers. Plans are regularly revised, using information that is more and more detailed over time and, therefore, less and less uncertain. In this way, well-founded plans can be developed, without the need for modification. Table 7.4 shows that we found that high-performing plants understand the value of the information necessary in order to achieve good planning.

Interestingly, we found that low-performing plants also understood the usefulness of computer systems for management of the flow of goods. Low-performing plants use materials resource planning (MRP) for order release and, to a lesser extent, for planning; but MRP is not used very much for shop floor control. In high-performing manufacturers, on the other hand, pull systems are the preferred method for shop floor control. Thus, high-performing plants base their materials flow management systems on just-in-time (JIT) principles, trying to integrate them with MRP systems so that the two systems do not conflict with each other (see Figure 7.5).

High performance manufacturers use MRP primarily for planning and, in particular, to plan open orders to suppliers. The plants that are faster in adding value to the product use planning systems for long-term decision making, for example, to make agreements with suppliers or to plan changes in mix, whereas they use pull systems for management of the flow through the work centers.

Table 7.4
Production Planning and Control

	p-value	High Performance	Low Performance
Stability of short-term production	0.00	3.17	2.78
Manufacturing plan	0.09	3.61	3.44
MRP adaptation to JIT	0.00	3.35	2.88

Figure 7.5
Uses of MRP and Pull System

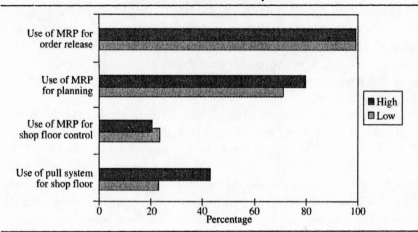

Information Systems for Integrated Supply Chain Management

High performance manufacturers tend to use integrated supply chain management for the planning and control of the total material flow, from the suppliers to the final customers. This increased integration modifies the set of constraints to which each member of the supply chain is subject and leads to optimization, both at the level of each individual supply chain member and at the level of the channel.

The management of product flow along the supply chain is complicated by the demand amplification effect, which can be traced back to the multiplicative and derivative effects inherent in the policies of reordering. Earlier information about future customer requests creates a climate of demand predictability, which makes it possible to organize the productive system appropriately. This information reduces uncertainty and makes it possible to eliminate some safety stock.

Early involvement of suppliers makes it possible for them to equip themselves for a ramp-type of channel, typical in the introduction of new products. A JIT supply relationship requires greater information exchange between a plant and its suppliers. In particular, early communication of a plant's productive needs allows its suppliers to obtain better time performance without being obliged to maintain costly stocks.

It is important to understand what the real trend in product demand is, independent from the reordering policies of single members of the supply chain. Open orders, for example, make it possible to reduce the disturbance provoked by the fluctuations in demand of a few large customers. With its biggest customers, a plant should agree on regular deliveries distributed over a period of time. The constant part of the demand is covered, leaving the management of variations to emergency planning. As a result, the cost of order management is reduced by 35 percent, while the average stock level is halved. High-performing plants, in fact, can be identified by their use of open orders with suppliers (see Figure 7.6).

It is important to consider the factors that encourage the passage to new forms of linkage between various members of the supply chain, foremost of which is information technology. Telecommunications opens new ways of doing business and modifies deeply engrained practices. It is therefore of great importance for a plant to not be excluded from the virtual network where more and more transactions take place. We found that high-performing plants are conscious of this transformation and use a greater extent of networking with supplier, customers, distributors, banks, and so forth (see Figure 7.7).

IT can greatly improve communication between the chain members, particularly in collaborative conditions, where information exchanges require greater frequency, significance, and timeliness. IT is the force behind improvement of the information flow and the sharing of information necessary for speeding up response of the entire supply chain. The improvements obtainable through an integrated information flow are considerable: the additional costs of managing a fluctuating production order rate can be reduced by more than 70 percent.

Figure 7.6
Mean Use of Open Orders

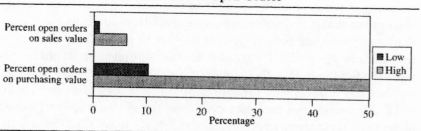

Figure 7.7
Mean Presence of EDI Linkages

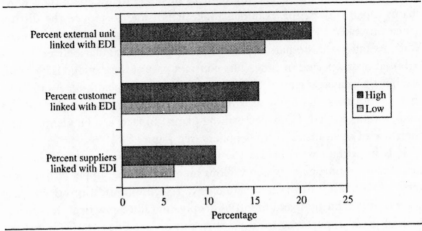

IT, therefore, can be important in the development of a distribution system that is fully integrated with the productive system along the entire supply chain. IT can be the hardware of a system that supplies high-level customer service and real feedback to the producer, allowing the system to function on the basis of customer requirements, without penalizing the requirement of the factory. However, in order to improve operating integration generally, the integration based on IT has to be preceded by organizational integration; a new culture oriented toward integration; and new strategies, structures, and organizational practices.

The use of telecommunications as a link between the suppliers and the factory makes it possible to achieve greater precision and speed in the communication of the necessary information for the control of the supply process, as well as continuous evaluation of the supply relationship. The matching of telecommunication systems to a supply organization through open orders makes it possible to achieve simplification of the routine activities of the supply process, greater stability in production, and greater flexibility on the part of the supplier to small upstream variations in volume. We found not only that high-performing plants use more networking, but also that a greater part of their business is done with the support of IT. By considering the order cycles, we found that high-performing plants exchange a greater percentage of orders electronically (Figure 7.8).

Figure 7.8
Mean Use of EDI for the Order Cycles

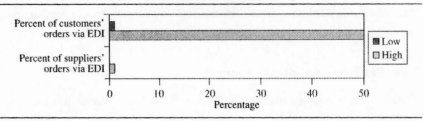

GOAL COMMUNICATION AND PERFORMANCE MEASUREMENT

It is important to consider communication of objectives and performance measurement systems because they are among the aspects that are most affected by changes induced by new production practices.

Communication of Goals

The need to define clear, shared, and nonconflicting objectives is important in HPM. In order to exploit synergies between functions, these objectives must be known to the personnel involved. High-performing plants try to communicate strategic objectives at various levels and to convert them into actions, in a cycle of communication and action. Therefore, the information system must ensure communication of the objectives and the information necessary for all people involved. Because they know the operating objectives of the business, empowered operators can make effective decisions using the information available to them. Our analysis (Table 7.5) shows that, in general, in high-performing plants, the essential characteristics of their strategies are communicated not only to their managers, but also to the

Table 7.5
Vertical Information Flows in the Decision-Making Process

	p-value	High Performance	Low Performance
Coordination with corporation	0.04	3.60	3.33
Communication of manufacturing strategy	0.00	3.73	3.34

operating levels, making them aware of the direction of the efforts being made by the plant. Thus, high-performing plants have a richer information flow of strategy information from the top down than do low-performing plants.

New Performance Measurement

Traditional systems for measuring performance, based mainly on financial measures, have been widely criticized because they encourage a misleading kind of control. In particular, they encourage only actions that lead to short-term results; they lack strategic focus; they encourage local optimization; they push managers to concentrate on the analysis of deviations rather than on improvement of the system; and they take too little account of customer satisfaction, rarely making comparisons with the competition.

A distinguishing characteristic of performance measurement systems in high performance manufacturers is that they base themselves on some type of dynamic performance measurement (see Table 7.6). This includes the use of measures that reveal both the evolution of phenomena and changes in measurement intensity. This is well adapted to the simultaneous pursuit of continuous improvement and attention to simplicity characteristics of high performance manufacturers. It would be absurd if in the face of simplification of the reporting systems connected to simplification of the production processes, a system that loads up managers with redundant, obsolete, or irrelevant information were to be introduced.

A second aspect that characterizes high performance manufacturers is the receivers of the information produced by the measurement system, including not only managers, but also operators and supervisors. Information feedback concerning costs, times, and quality arrives at lower levels, with the aim of providing useful information for the work with which these people are involved. This is further reinforced by making feedback

Table 7.6
Performance Measurement in High Performance Manufacturing

	p-value	High Performance	Low Performance
Dynamic performances measure	0.01	3.78	3.44
Multidimensional performance feedback	0.04	3.60	3.42
Advanced accounting systems	0.77	3.25	3.22

highly visible and accessible. Think, for example, of the huge charts exhibited in the production departments of high performance manufacturers, charts that report performance measurement results.

We are certainly not advocating the abandonment of traditional accounting systems or attention to cost variations. Traditional accounting systems remain important. However, the nontraditional accounting systems used to supplement them in high performance manufacturers make a greater contribution in terms of the identification of cost drivers and the elimination of activities that do not add value.

SOME FINAL CONSIDERATIONS

More Appropriate Choice and Use of IT, Rather than Greater Expenses

When comparing high- and low-performing plants, we noted that there is no difference between them in terms of information system expenses or articulation (Table 7.7). Likewise, hardware architecture, including both the size and the modernity of hardware, is not significantly different between high- and low-performing plants. Software architecture, including the presence of applications such as CAD, CAE, CAPP, and MRP and the use of the database applications, is not particularly differentiated either. The telecommunications architecture (the presence of communication networks inside the plant and as electronic links with external units) does not seem to be particularly different between high- and low-performing plants.

However, we did find some differences between high- and low-performing plants, for example, in the quantity of orders exchanged via

Table 7.7
IT Expenses

	p-value	High Performance	Low Performance
IT expenses on sales	0.34	1%	1.1%
Incidences of IT maintenances costs on IT expenses	0.86	11%	12%
Incidences of telecommunication costs on IT expenses	0.41	6.7%	10%

EDI; in the percentage of design data electronically exchanged with customers, suppliers, and other plant units; in the use of MRP for the planning of open orders; in the use of databases for the collection of data concerning quality; and in the use of automatic inspection systems. Thus, the difference between high-performing plants and the others from an information systems point of view lies in *the way information systems are inserted in the processes,* that is, in how these processes exploit the potential that they offer. This observation is reinforced by managers' opinions concerning the benefits to be gained from the use of information systems in the plant (Table 7.8). Both high-performing-plant managers and those in other plants say that they have obtained considerable benefits from the introduction of information systems into their plants. The former group declare that they have had greater benefits compared to the latter group, however.

The message that emerges is not to spend less on IT. The message is that *spending on IT does not assure improvements and results.* It is important *to choose and to use IT in a better way.* This implies a better linkage between knowledge of the operating processes and IT offerings. Better choice and better use, in some cases, could mean rethinking the kind of IT that will be used. In comparison with a few years ago, today there is a greater variety of choices in IT, and, in particular, there are many solutions that can be adapted to the particular situation with limited costs and greater maintenance support over time.

However, our data can be somewhat misleading because the driver for the IT expenses is complexity. IT expenses increase as complexity increases. Therefore, to be rigorous, one should compare plants characterized by the same complexity level: similar product complexity (an airplane is

Table 7.8
Perceived Benefits of IT Introduction

	p-value	High Performance	Low Performance
Benefits of IT (global)	0.00	3.73	3.61
IT support in cost and inventory reduction	0.01	4.00	3.67
IT support in customer's and supplier's management	0.02	3.67	3.50
IT support in time reduction	0.01	3.73	3.47
IT support in service and quality improvement	0.02	3.56	3.33

far more complex than a pencil), similar process complexity, and similar supply networks. We did this and found that, given the same level of complexity, the plant that spends more on IT is usually *not* the higher performer; very often it spends more to face managerial problems that can be solved in an easier way, for example, with better management or with a better overall organization.

Complexity and the Level of Computerization

Information technology is increasingly used in all sectors and in all kind of activities. Differences can be found among sectors, dimensions, products, and countries, but many of these differences have changed rapidly as time has passed. The main difference that remains over time is the level of complexity, in terms of product/process complexity and complexity of the overall organization, and the level of computerization. For example, the machinery sector, which is characterized by less repetitiveness, complex exchanges of information with customers, and more sporadic orders from customers, uses less on-line quality control, EDI, and external telecommunications. It also uses fewer open orders and combines MRP with the pull system. On the other extreme, auto suppliers, characterized by greater repetitiveness and by continuous supply linkages with their customers and from their suppliers, use more electronic linkages with suppliers and customers both for the order cycle and for design activities. They also use more open orders with customers and suppliers and use a combination of MRP and the pull system.

Great differences can be found between small-to-medium-size manufacturers and larger manufacturers. Larger manufacturers tend to reduce complexity by externalizing operations to specialized suppliers or subcontractors. Smaller manufacturers face limited complexity because they are very skilled in the particular phase in which they operate and have a more focused production system. This situation facilitates their shop floor control and makes a complex shop-floor-control-information system unnecessary.

The way that excellent manufacturers manage their operations leads to fewer requests to information systems (IS). The choices that they make in their relationship with the supply chain network can greatly reduce the complexity they have to face. The externalization of production phases not only has reduced internal complexity but, in many cases, also has reduced shop floor control complexity at the network level, since the specialized supplier has a more focused production system. The externalization of production phases, in combination with better specific skills which allow machines to be used in a more flexible way, leads to less complex shop floor control and therefore to simpler information systems. On the other hand, it increases the necessity to have IS support to manage flows at the supply chain network level.

IS Design Tailored to Process Performance

In HPM, great emphasis is placed on the value of information and on making this information worthwhile. For this reason, information is viewed as a *resource* for keeping operations under control and for improving them. The importance of information means that a plant that wishes to become a high-performing plant may need to modify both its information flows and its ways of using information technology. HPM requires more horizontal information flows, different vertical information flows, and greater information exchange with customers and suppliers. Horizontal information flows increase to ensure process coordination and problem solving: departments in the plant communicate frequently with each other, and supervisors encourage people who work for them to exchange opinions and ideas and frequently hold group meetings where the people who work for them can really discuss things together. Vertical information flows consider the problem of communication for achieving improvement and for direction setting, in depth. Therefore, suggestions, feedback on suggestions, feedback on appropriateness of work, and goal communication are some of the contents of vertical communication. External information flows increase greatly with reference to all the main processes: product/process design, transformation, and management of the physical flow. This is related to efforts to reduce uncertainty, to stabilize flows, to anticipate problems, and to improve operations at the supply chain level because HPM operates on the premise that the time, quality, and cost performance that the final customer experiences is the result of the time, quality, and cost performance of each member of the supply chain.

High-performing plants support managers and operators in their everyday working activities by providing information support. They are very careful in the identification of the most suitable applications, covering those aspects that have the most influence on the global performance of the processes. This approach leads them to introduce applications that are more closely connected to the processes. *They use IT as a tool that allows a redefinition of the processes and not a tool to which processes must be adapted.* CAD, CAE, CAPP, databases, and telecommunications are instruments for redesign and product/process development, challenging the traditional phase linkages both inside and outside the plant with customers and suppliers. The results come from the new process and only in a limited way from the use of a specific technology. MRP, EDI, and bar coding are only instruments that allow better channel coordination in the management of the flow of goods.

It is the giving of a new shape to the process that leads to superior performance: for example, redesign of the supply network provides performance gains. Without IT, this new process configuration is almost impossible, but a plant can introduce these technologies without reaching the top level of process configuration. Numerically controlled machines, robotics, automatic inspection, databases, and so forth, make greater control of the transformation process possible. However, the information gathered by them has to be used for improvement. IT should be considered as a *tool to enable and support the adoption of high performance production practices or as an ally in the redesign of production processes in order to make them more efficient.*

The purchasing and installation of IT is not sufficient to obtain some competitive advantage; rather, a lot depends on its actual use, on the attitude of managers and operators to the matter, and on the vision that considers information instruments as nonexclusive levers for process improvement.

As Tony Salvatori ponders this information, he realizes that manufacturing excellence may be possible through selective investments in information systems. Because his plant is already networked and has implemented many HPM processes, such as a strong network of communications with suppliers and customers, Tony is cautiously confident that he will be able to use limited investments in IT as a lever to get the greatest benefit out of his plant's evolving processes and management approaches. In his plant IT will be used to support and develop the processes implemented, and not the other way around.

REFERENCES

Davenport, T.H. *Process Innovation: Reengineering Work Through Information Technology.* Boston: Harvard Business School Press, 1993.

Galbraith, J.R. *Designing Complex Organizations.* Reading, MA: Addison-Wesley, 1973.

Schonberger, R.J. *World Class Manufacturing: The Lessons of Simplicity Applied.* New York: Free Press, 1986.

CHAPTER 8

JIT MANUFACTURING: DEVELOPMENT OF INFRASTRUCTURE LINKAGES

SADAO SAKAKIBARA, BARBARA B. FLYNN, and
ALBERTO DE TONI

Although *just-in-time* practices (JIT) have been discussed and used by manufacturers around the world since the early 1980s, there is still little agreement on what is meant by the term. To some, JIT is a set of narrowly defined practices designed to minimize inventory or to schedule production. To others, it is a more comprehensive approach or even a philosophy of manufacturing. In this chapter, we will use our data on manufacturers in five countries to examine both the narrow and the broad definitions of JIT. We will also look at differences in approaches to JIT by country, by industry and by volume level in order to examine alternative paths to JIT performance.

The HPM perspective of JIT is broad, viewing JIT as a comprehensive approach to continuous manufacturing improvement, based on the notion of eliminating waste in all phases of the manufacturing process. If adopted properly, JIT functions as an engine for a plant to move forward and to create an evolving organization with the capability of adapting to the changing business environment.

Failures in implementing JIT often come from a focus that is too narrow, leading to results that are much less than originally anticipated. We have found that the plants with the best JIT performance have an infrastructure of practices that support JIT. This illustrates the notion of levered linkages. The plants with the best quality management, human resource development, and strategic management practices have an infrastructure foundation for building effective JIT performance. Thus, a

141

narrow focus, such as operational efficiency, should not be the goal in implementing JIT. JIT affects a much wider range of organizational goals, well beyond just the practices on the shop floor.

ORIGIN OF THE CONCEPT AND BACKGROUND

History of JIT

Although the first articles about JIT written in English did not appear until the late 1970s, the original concept of JIT can be traced much farther back. It was in 1936 that Kiichiro Toyoda, the son of Toyota founder Sakichi Toyoda and president of Toyota from 1941 to 1950, first discussed the concept with manufacturing managers at his Kariya assembly plant. He hung the words "Just-in-Time," in the Japanese language, on the walls of his office and tried to spread the concept to all employees in the plant. In fact, he often wandered around the shop floor by himself, insisting that workers should only produce what was needed, when it was needed, and in the amount needed.

The concept of JIT did not originate in Kiichiro Toyoda's manufacturing experience. He conceived of the basic idea of JIT during a visit to the United States when he visited several factories, including automobile plants. But the idea of JIT was not derived from his plant visits. Rather, it came from Toyoda's visits to U.S. supermarkets, where he observed that customers tossed merchandise into their carts and the store simply replaced the products that were sold. This "take away and then replace" process gave him the idea of the pull system. Using a pull system, each work center produces exactly the amount that is taken away or used by the next process, no more and no less. By insisting on this approach in his factory, he believed that waste in the manufacturing process could be eliminated.

However, if the title "father of JIT" could be conferred on anyone, it probably should go to Taiichi Ohno. Born in Japan in 1912, he graduated as a mechanical engineer in 1932. He went to work for Toyota in the production section and, as time passed, was promoted to positions of greater responsibility until 1985, when he became executive vice president of the company. When he retired, he wrote a book entitled *Toyota Production System,* which summarized his experiences.

In describing the production logic developed over the decades in the Japanese automobile industry, Ohno described himself as the most "Fordist" of his contemporaries. Henry Ford's dream was to attain *totally synchronous* production, in which not only assembly took place in the "high line," but also the entire production of components upstream would be synchronized with the incessant rhythm of the assembly line.

However, this vision was never realized by Ford because of the technological limits of time and competition imposed by General Motors. The proliferation of models away from the legendary "model T"—famous because it was available in "any color as long as it is black"—impeded the realization of Ford's dream. The complexity of the production cycle and the diversification of models led to the division of production into various upstream departments, decoupled by intermediate warehouses. Within the same department, space for inventory buffers had to be provided between one operation and the next in order to allow the necessary processing without having to continuously change over the machinery. Thus, the factory could not become the conduit where the raw material synchronously entered at one end while finished product came out the other.

The challenge that had been lost by Ford was taken up by Ohno who, by introducing innovative organizational methods and management, dealt with the complex problem of changing intermittent upstream manufacturing into repetitive production, even for different models. Ford's vision of total flow, which was abandoned at the beginning of the 1920s, was relaunched by Ohno. In this sense, it is only proper to consider Ohno as the true continuer and accomplisher of the pure Ford vision. His approach can, with good reason, be defined as a type of "Fordism surpassing Ford."

The Concept of JIT

If the principle of "totally synchronous production" is considered the point of similarity between Ford's vision and JIT, then the principle of "adaptive synchronism" represents the first point of departure. With the introduction of "pull" logic—the method by which the downstream departments trigger upstream production according to the required mix—Ohno reversed the "push" logic of the shop floor. As Ohno remembers, Kiichiro Toyoda returned home with the idea that it was the right time for Japan to produce "just in time" because the Japanese market was not big enough to justify mass production, as he had seen in Ford's factory in the United States.

"Toyotism," Ohno stated, "is the antithesis of the American system of mass production." "The Toyota production system was conceived for slow or zero growth conditions," and "the system encourages production to get closer to the market." If in Fordism the important thing is to keep up the productive rhythm, then in JIT it is fundamental to produce synchronously, but in the quantities demanded by the market. This is contrary to the idea that it is right to decouple production and demand by having stocks of finished products so as to protect production from market turbulence. Thus, while traditional thinking seeks to protect the production processes from the ups and downs of the market, JIT logic recognizes that market ups and downs are inevitable, striving to synchronize the production process with them.

Moving to synchronized production implies a second great break, after "adaptive synchronism," with Ford's vision: it is that related to human resources. Rather than the traditional antagonistic relationship between management and the workforce, the workforce is perceived as a resource in JIT. Adaptive synchronism presupposes the participation of the entire labor force in order to keep demand and production in "pull." As summed up by Bonazzi (1995), a plant resembles a crystal tube through which material flows rapidly. The flow of materials is very fragile, like glass. The elimination of the work-in-process (WIP) inventory and of every other slack resource strips away its defenses, making it very vulnerable.

Ohno mentioned in his book that, up to the time of the oil crisis of the early 1970s, there was not much interest in the Toyota system. He was willing to share his knowledge and, in fact, talked about the Toyota production system to his Japanese industry colleagues, but their interest was almost nil. However, the oil shocks in the 1970s changed the attitude of Japanese manufacturers toward the Toyota system. Many manufacturers were having difficulties making profits; but even during this difficult period, Toyota maintained profitable operations. The gap between Toyota and other automobile manufacturing firms widened even farther from the middle to late 1970s. It was in the 1970s, then, that other Japanese manufacturers began to learn about the Toyota system, and its concept rapidly spread all over Japan during the 1970s and 1980s.

As the Japanese automobile industry expanded and a flood of Japanese cars started to appear abroad, automakers in other parts of the world began to pay attention to the Japanese ability to make cars with high quality at relatively low costs. In 1977, a group of Toyota managers published an article about the Toyota production system, the first article about JIT to be published in English. This was followed by Monden's series of articles in 1981 and his book, entitled *Toyota Production System*, in 1983. Around the same time, Schonberger's book *Japanese Manufacturing Techniques* was published, based on his study of a Japanese transplant (Kawasaki) in Nebraska. Because of this early literature and the success of Japanese automakers, manufacturers worldwide started to study the JIT concept and to implement this approach in their operations. The rest is history.

STATE OF THE ART IN JIT

JIT is now implemented widely, in many parts of the world. In fact, we found that the extent of JIT usage in most of the countries in our data set was similar to the extent of JIT usage in Japan. Figure 8.1 shows that plants in Germany and the United States had comparable levels of JIT use, followed closely by plants in Italy. We found, however, that the plants in the United Kingdom did not use JIT to nearly the same extent as the plants that we studied in the other countries.

We used the plants that we studied to determine patterns of JIT implementation. Not surprisingly, we found that JIT was implemented much earlier in Japan than in the other countries we examined. Figure 8.2 shows the range of JIT adoption dates among the plants in our sample. The length of the vertical lines show the range between the date when the first plant in that country in our sample adopted JIT and the date when the last plant in that country in our sample adopted JIT. For example, the earliest Japanese plant began using JIT in 1960, while the latest began using JIT in 1992. Thus, it is reasonable to assume that JIT is at a much more mature level in Japanese plants. In contrast, in the United States, the earliest JIT adopter began in

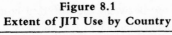

Figure 8.1
Extent of JIT Use by Country

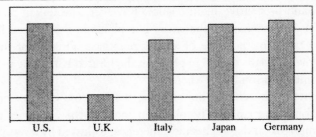

Figure 8.2
Range of Start of JIT Implementation by Country

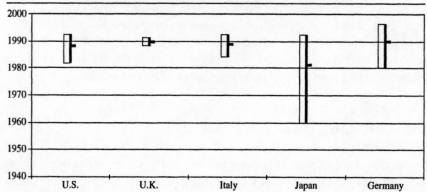

1982. This is consistent with the earliest discussions of JIT use in the United States in the literature, indicating that, although it was 22 years later than the earliest Japanese adoption, this plant was a real JIT pioneer, by U.S. standards. Figure 8.2 shows that the earliest plants in Germany, the United States, and Italy began using JIT at around the same time. The earliest adopters lagged the other countries, not beginning JIT use until 1989.

The small horizontal lines indicate the average point at which JIT was implemented in each of the countries we studied. The average year of JIT implementation was 1981 in Japan, before the time of first implementation in the other countries we studied. In the other countries we studied, the average year of adoption of JIT was 1988 to 1990.

We also examined variations in the way JIT was implemented in various countries. Figure 8.3 shows only the practices in which we found significant differences in use between the different countries. Japanese plants had the best average level of practice in JIT delivery by suppliers, setup-time reduction, and preventive maintenance; however, the Italian plants were strongest in daily schedule adherence. The plants in the United Kingdom were the weakest in their use of all JIT practices. This is not surprising, given the low emphasis on JIT in plants in the United Kingdom and the relatively recent adoption of JIT there.

The plants in the United States made the best use of most of the JIT practices, following Japan. The plants in most of the countries put their primary emphasis on daily schedule adherence, followed by a roughly equal

Figure 8.3
JIT Practices by Country

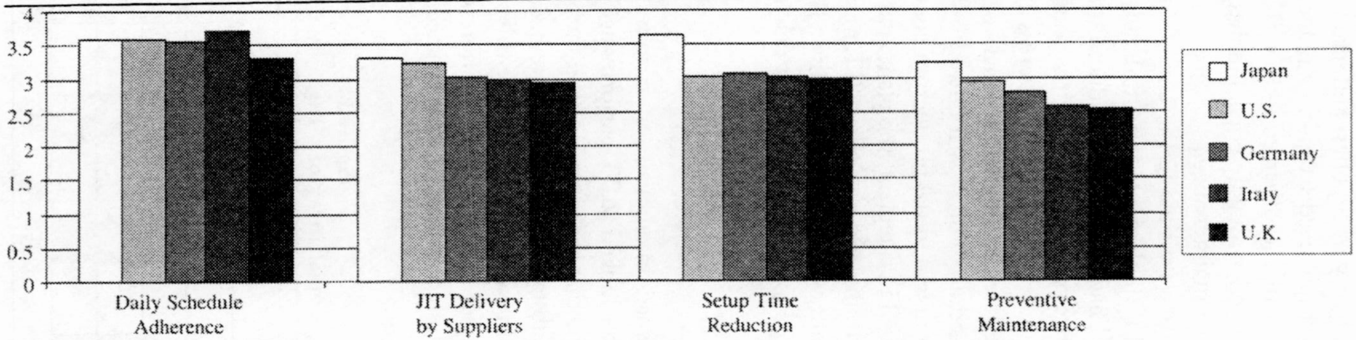

Legend:
- Japan
- U.S.
- Germany
- Italy
- U.K.

Categories: Daily Schedule Adherence, JIT Delivery by Suppliers, Setup Time Reduction, Preventive Maintenance

Y-axis: 0, 0.5, 1, 1.5, 2, 2.5, 3, 3.5, 4

147

emphasis on JIT delivery by suppliers and setup-time reduction. However, the plants in Japan emphasized setup-time reduction to a much greater extent than the plants in the other countries. This was followed by daily schedule adherence and a roughly equal emphasis on JIT delivery by suppliers and preventive maintenance.

One of the plants we visited in the United States demonstrates the approach taken to JIT implementation. It already had a very good quality system in place and had significantly reduced machine downtime through superior maintenance procedures. Next, the plant started with reduction in setup times of all major equipment and cross training of the workforce. Once this was completed, it could go to small lot sizes and a repetitive master production schedule. This was followed by implementation of a pull system of production and visual production control by the workers. Once JIT was working well in the plant, the plant approached its suppliers to explain what it was doing and to request JIT implementation by suppliers. Eventually, the plant was also able to deliver JIT to its customers, using significantly reduced lot sizes and frequent deliveries.

Performance

How do the differences in emphasis on various JIT practices relate to performance? When we looked at JIT performance in terms of annual inventory turns, we found that the plants in Japan outperformed plants in the rest of the world by a substantial margin. Figure 8.4 shows that Japanese plants averaged 21 annual inventory turns, while this ratio was quite a bit lower in the other countries. Thus, although different JIT practices have been followed in the different countries, the Japanese plants have achieved the

Figure 8.4
Annual Inventory Turns by Country

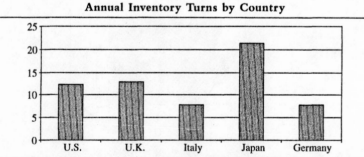

best performance. This shows that, at least in the case of JIT, there may be some paths to performance that are superior to others. However, although JIT has not been widely implemented in the United Kingdom and JIT practices are weakest there, plants in United Kingdom were second only to Japan in annual inventory turns, at an average of 13 per year. This provides support for the concept of different paths to performance, in a broader sense. Many plants have chosen not to aggressively implement JIT; however, they have been able to achieve competitive levels of inventory turns by other means.

We also looked at the use of JIT practices by volume and by industry. We found, as expected, that the higher volume (large batch and repetitive/line flow) plants were stronger in their use of all the JIT practices, compared with one-of-a-kind and small-batch manufacturers. When we looked at the use of JIT practices by industry, we found that there were no differences between the three industries we studied in terms of their use of practices related to daily schedule adherence, setup-time reduction, preventive maintenance, and material handling simplicity. However, we did find significant differences in the use of practices related to JIT deliveries by suppliers and MRP adaptation to JIT. Plants in the transportation components industry led the way in their use of each of these practices. This is consistent with the historical dominance of the automotive industry in the use of JIT and the push of JIT practices to suppliers of large automobile plants. The transportation components industry was followed by the electronics industry in use of practices related to JIT deliveries by suppliers and MRP adaptation to JIT. Plants in the machinery industry were the weakest in the usage of these two practices.

JIT IS MUCH MORE THAN JUST AN OPERATIONAL TOOL

As discussed previously, JIT failures often come from efforts that are too narrowly focused. The HPM approach stresses the importance of the linkages between JIT practices and the infrastructure of practices that supports their use. Even in the early JIT literature, there was an awareness of the vital contribution of a supportive infrastructure to the success of JIT. For example, one of the earliest English discussions of JIT described the Toyota system as having two components, working hand in hand. The first was the *JIT production system*, described as "only the necessary products, at the necessary time, in the necessary quantity." This is the well-known perception of what some think is the entirety of JIT. However, the second component, described as equally important by the early Japanese authors, was Toyota's *respect-for-humans system,* which focuses on active employee participation, elimination of wasted movement by workers, and self-display of workers' capabilities by entrusting them with greater responsibility and authority. One of the goals of a JIT system is raising the level of shop floor control through decentralization, by giving workers on the floor the function of production control and inventory control.

In our previous research, we found a strong indication that the level of JIT implementation is highly associated with many other subsystems or functions of the organization, including the areas of manufacturing strategy, quality management, and human resource development. Figure 8.5 shows the overall relationship between JIT and infrastructure practices. Although the primary contributor to JIT performance (inventory turns, cycle time, etc.) is expected to be a strong core of JIT practices, it is also expected that quality management, human resource development, and

Figure 8.5
Linkage between Core JIT Practices and Infrastructure

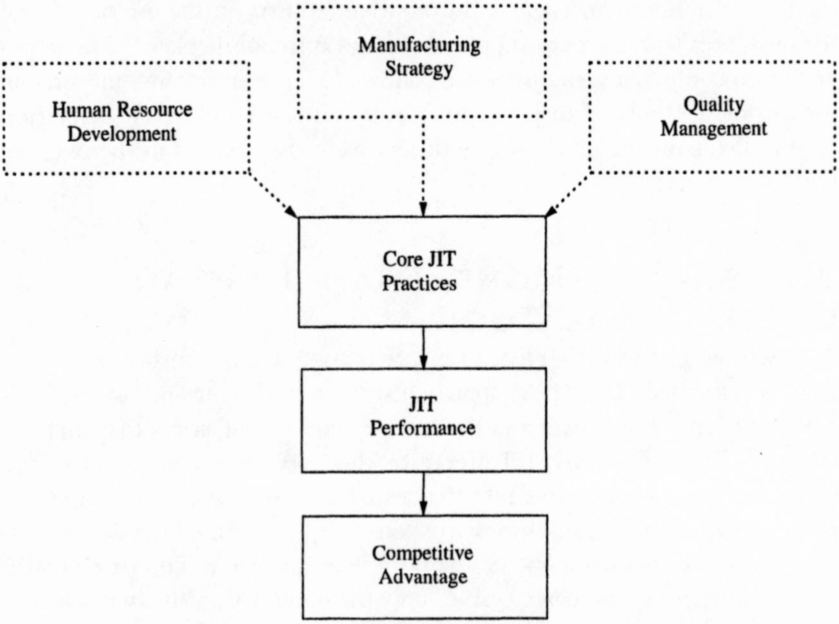

manufacturing strategy will support the use of JIT practices. Thus, all other factors being equal, plants that are stronger in these three areas will have better JIT performance.

Infrastructure for JIT

In examining the relationship between JIT and infrastructure practices, we concentrated on JIT linkages with three primary areas of infrastructure, although other areas contribute as well.

Quality Management Practices.   The linkage between quality management and JIT has been well understood since early JIT development at Toyota. Quality management activities support JIT through the establishment of a process that is in control. This facilitates the development of an unhampered flow of goods through the process, allowing the reduction or the elimination of buffer inventories. The provision of accurate and timely feedback about the manufacturing process permits shop floor personnel to detect, diagnose, and remedy process problems as they occur. Quality management also improves JIT performance through process variance reduction, which leads to reduced rework time. This is not a one-sided relationship, however. JIT improves quality performance through exposure of problems that can be solved using quality management tools and through improved process feedback. Thus, JIT and quality management are linked synergistically, in a mutually supportive fashion. Effective implementation of JIT is not possible without the necessary support of quality management, and quality management functions more effectively when JIT's inventory reduction efforts help to expose problems in the process.

We examined the role of quality management practices by dividing the plants in our data set into those plants that did not use JIT, those that used JIT in a limited sense (only some JIT practices or JIT only in part of the plant), and those that had fully implemented JIT. We found that there was a difference among these plants on four, key, quality management practices: (1) physical working conditions, (2) feedback, (3) top management support for quality, and (4) rewards for quality (see Figure 8.6).

The plants that had fully implemented JIT were more likely to have taken steps to organize the workplace and maintain it in order to help employees accomplish their jobs more efficiently and to instill a sense of pride in their workplace. Examples of practices related to physical working conditions include the use of pegboards in work centers with painted

Figure 8.6
Quality Management Infrastructure by Extent of JIT Implementation

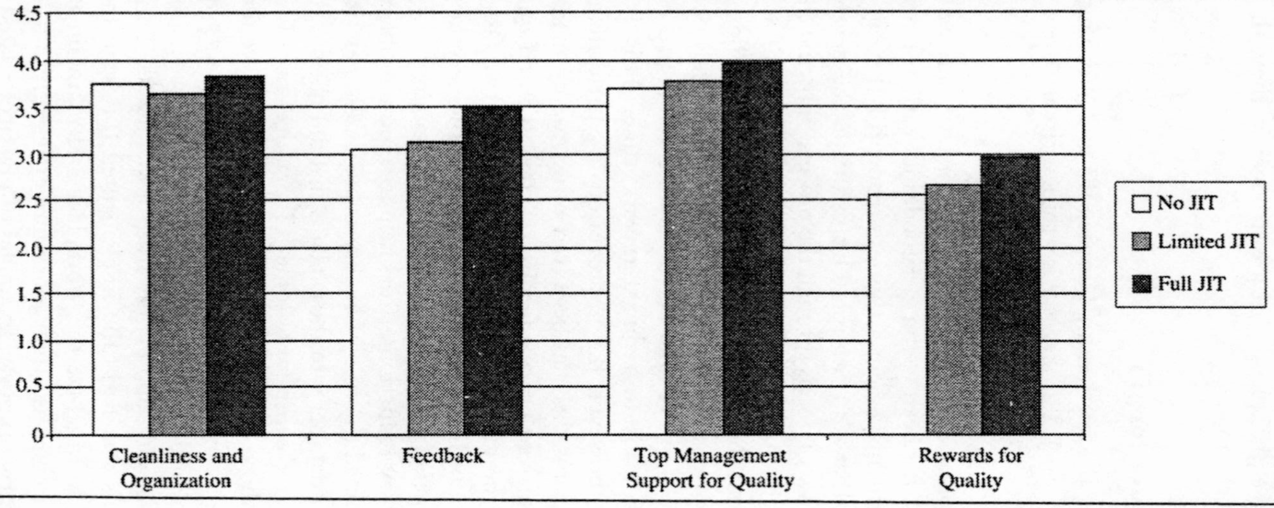

No JIT
Limited JIT
Full JIT

Rewards for Quality

Top Management Support for Quality

Feedback

Cleanliness and Organization

4.5
4.0
3.5
3.0
2.5
2.0
1.5
1.0
0.5
0

silhouettes of tools on them to show where particular tools should be kept. Neat and clean workplaces are also important.

We also found that feedback was stronger in the plants that had fully implemented JIT. It was more common to see charts posted that indicated defect rates, schedule compliance, machine breakdowns, and so forth, as well as information provided to individual employees about their own performance. This indicates that these plants provided shop floor personnel with information regarding their performance in a timely and useful manner. Information about individual and plant performance is important in keeping JIT flows operating smoothly.

Top management support for quality was greater in the plants that fully implemented JIT, as exemplified by top management commitment and personal involvement in pursuing continuous improvement. By setting an example, top managers encourage everyone in the plant to be committed to quality improvement. Top management commitment to quality was also illustrated by the greater use of rewards for quality in the plants that have fully implemented JIT. This included rewards for both workers and supervisors, individual and group incentives, and bonus systems, as well as nonfinancial incentives.

Strength in these quality management practices combines to form an infrastructure for JIT. As quality management improves, it is much less risky to reduce inventory levels in JIT, knowing that there is less need for inventory buffers to compensate for defective products. In addition, the process will flow more smoothly when it is not encumbered by quality problems.

Human Resource Development Practices.   Human resource development is the second important element of the infrastructure linked to JIT. Its relationship with JIT includes selection and compensation policies that develop team players, who contribute to group problem-solving efforts. Training programs and personnel policies are important in developing flexible employees who are capable of shifting to the place in the process where they are most needed in order to facilitate a smooth production flow. Small-group problem-solving activities provide solutions for production-related problems, leading to improvements in waste reduction and manufacturing cycle time. JIT is also involved with the reallocation of decision-making authority within a plant and leads to development of a human resource management system that helps to develop a decentralized organizational structure. Such a structure will help in the development of a flexible, informed, and participative workforce capable of solving problems as they arise through coordinated activities.

Figure 8.7 shows the use of human resource development practices in plants that use JIT to either a full or limited extent, as well as in plants that do not use JIT. We found that seven workforce practices were superior in the plants that had fully implemented JIT. These included practices related to (1) decentralization of authority, (2) coordination of decision making, (3) implementation of employee suggestions, (4) recruiting and selection, (5) small-group problem solving, (6) supervisory interaction facilitation, and (7) multifunctional employees.

The plants that had fully implemented JIT were more likely to use a decentralized organizational structure, providing employees with a greater amount of autonomy in decision making. This allowed the plants to keep the line running because employees could make decisions about problems quickly, without having to refer every issue to a higher authority for approval. Superior practices related to coordination of decision making led to a perception of better cooperation and communication within the plant. Employees perceived that the people in the plant work well together and communicate frequently, rather than being in a state of conflict. In a JIT environment, cooperation and communication are critical to the rapid decision making necessary for smooth flows between operations. This decentralized and cooperative environment is exemplified by the higher level of implementation of employee suggestions in the plants with better JIT performance. This indicates that management takes employee suggestions about improvements to products and processes seriously and that employees are encouraged to make suggestions.

The plants that had fully implemented JIT used better recruiting and selection practices, which were targeted at selecting employees who have strong teamwork and problem-solving skills and whose personal values match the organization's values. Employees who can work well in small groups and who have an aptitude for problem solving are important to minimizing waste and keeping the process flowing in a JIT environment. Small-group problem-solving was, indeed, better in the plants that had fully implemented JIT. Teams were effectively used for continuous improvement on the shop floor. The plants that had fully implemented JIT used supervisors to encourage this cooperative, decentralized, problem-solving atmosphere. The supervisors helped workers to function as a team, expressing their opinions and cooperating with each other to improve production and process flows. Supervisors also organized frequent small-group

Figure 8.7
Workforce Management Infrastructure by Extent of JIT Implementation

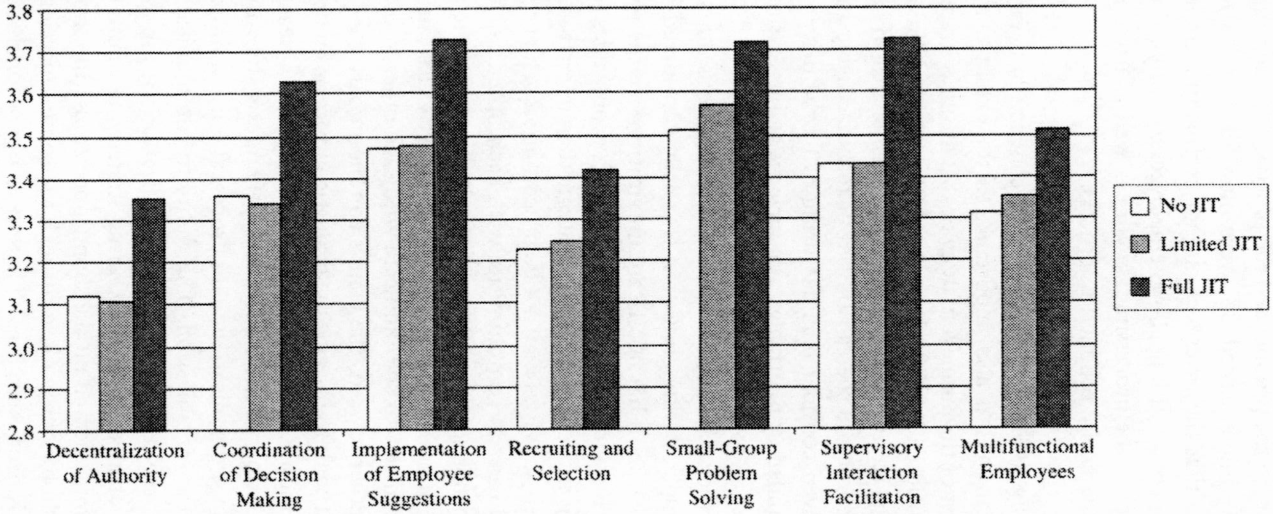

Legend:
- ☐ No JIT
- ▨ Limited JIT
- ■ Full JIT

Categories (left to right):
- Decentralization of Authority
- Coordination of Decision Making
- Implementation of Employee Suggestions
- Recruiting and Selection
- Small-Group Problem Solving
- Supervisory Interaction Facilitation
- Multifunctional Employees

Y-axis: 2.8, 2.9, 3.0, 3.1, 3.2, 3.3, 3.4, 3.5, 3.6, 3.7, 3.8

meetings to facilitate discussions among the people they supervised about improvements to the product and the process. We also found that the plants that had fully implemented JIT were more likely to provide training for employees so that they were capable of performing multiple tasks in the process. Such cross training allows employees to easily fill in for absent employees, keeping the processing flowing as well as having a better understanding of the entire production process.

Manufacturing Strategy Practices. The connection between manufacturing strategy and JIT is also important. To the extent that JIT becomes an important part of the business strategy or that business strategy is built on the capabilities offered by JIT, there should be a strong linkage between the manufacturing strategy and JIT practices. Well-managed organizations progressively incorporate the capabilities of manufacturing into their business strategy or even consider manufacturing to be the driver of their business strategy. Manufacturing strategy also includes an integration of strategic and functional decisions.

Consider the strategic role of Kaizen (continuous improvement) activities, for example. Although Kaizen is usually considered as strictly an operational function, it has significant implications for the implementation of strategy. For instance, although many Japanese companies encourage Kaizen activities, this move is often more strategic than operational. The true purpose of Kaizen activities may not be process improvement *per se,* but rather forcing employees to communicate with each other. For this reason, even ideas that do not seem to generate much direct benefit are often implemented. The actions associated with Kaizen activities cause workers, managers, and engineers to communicate with each other, providing the vehicle to work together towards the plant's common goals.

It is not difficult to imagine the dramatic impact on an organization where over two million Kaizen ideas are implemented each year, such as Toyota. The accumulation of those tiny steps generates a dynamic, learning, and ready-to-change organization. This process also has an impact on the strategic aspects of an organization. The ability to communicate provides information about the true picture of an organization for top management. The result is that changes in manufacturing strategy are more likely to be welcomed and implemented on the shop floor because the strategy reflects the voice of the shop floor workers through the communication channels developed using Kaizen activities. Therefore, operational activities, such as Kaizen, not only contribute to JIT and quality management, but also form a linkage between strategy and operations. Because Kaizen is part of strategy, top management should be responsible. At Toyota, Kaizen activities are coordinated by a senior vice president.

We compared the strategic management practices between the plants that had full and limited JIT implementation and those that had no JIT implementation. We found significant differences in four practices: (1) anticipation of new technologies, (2) communication of manufacturing strategy, (3) the use of a formal strategic planning process, and (4) practices related to the manufacturing-business strategy linkage (see Figure 8.8).

Plants that anticipate new technologies are strategically well prepared in advance of technological breakthroughs, constantly scanning the environment, searching for technological breakthroughs that might be relevant to their operations. When potential technological breakthroughs are detected, these plants learn as much as then can about them in order to evaluate whether they might be appropriate to pursue. Thus, these plants are forward looking, considering which technologies might be relevant to them in the future and acquiring and mastering new manufacturing capabilities in advance of their needs. Because they are looking toward the next generation of technology, they will find it simpler to incorporate changes in the process when needed, lessening the disruption to the process. In this way, they are able to aggressively pursue new technologies when the time comes, leaving their competitors to play catch-up.

The plants that had fully implemented JIT used a more formal strategic planning process, frequently updating strategic plans. They were more likely to have a written mission, long-range goals, and strategies for improvement. They also had strategic business plans that were linked to the strategic manufacturing plan. Plants that have an explicit manufacturing strategy use their strength in manufacturing to support the overall business strategy and to support JIT. Potential manufacturing investments are screened for consistency with the business strategy, and the business strategy is translated into its business implications. Thus, JIT is viewed as an element in achieving the strategic goals of the organization, and it is accorded a high priority.

JIT Pushes Other Practices to Higher Levels.   We have discussed the way that quality management, human resource development, and strategic management form an infrastructure that helps support JIT performance. However, this is not a one-sided relationship. Strength in JIT pushes performance in other areas to higher levels, as well. For example, one outcome of strong JIT performance is reduced inventory levels. It is true that reducing the level

Figure 8.8
Strategic Management Infrastructure by Extent of JIT Implementation

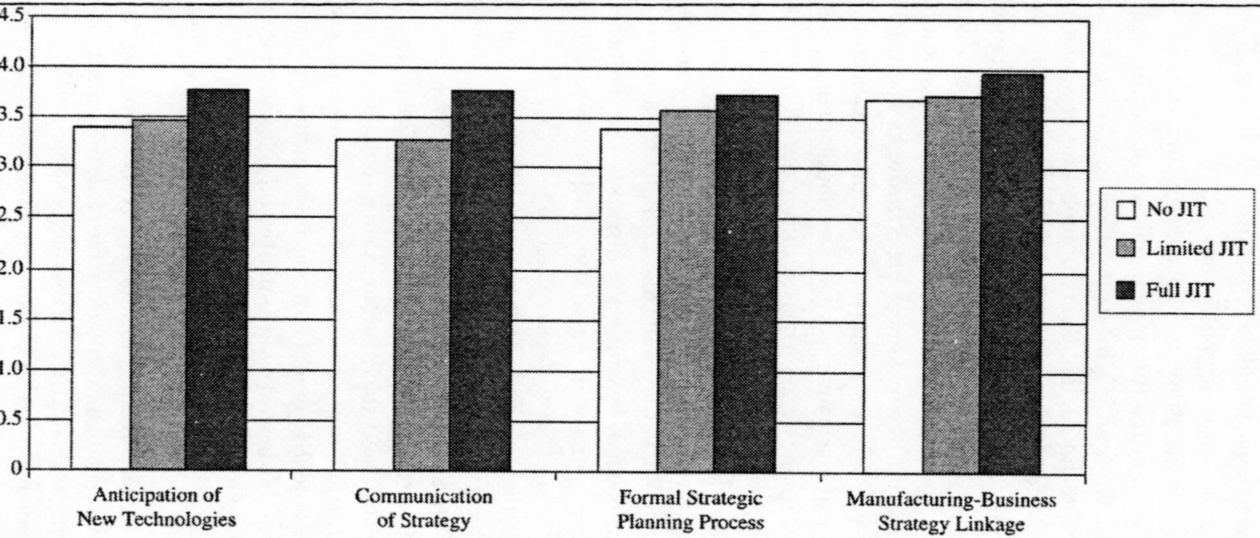

158

of inventory through the use of JIT practices improves manufacturing cost. However, the true purpose of reducing inventory in JIT is forcing other functions of the organization to excel. Reducing inventory exposes problems, pushing workers to reduce defects and thus improves quality performance. Inventory reduction also forces the organization to develop a more effective supply chain network, which involves strategic decisions by top management. Inventory reduction may also lead to development of better approaches to training employees in the basic concepts of quality management because there are no longer inventory buffers to compensate for errors. It will also likely encourage people in different functions on the shop floor to communicate and to coordinate their decisions and actions because decisions will need to be made quickly in the absence of inventory.

Thus, every time the targeted level of JIT becomes more ambitious, so must the targeted levels of the infrastructure practices, to capitalize on levered linkages. Otherwise, JIT will not function effectively. A weakness in one JIT infrastructure area will become a trapped linkage. Thus, if JIT is implemented without strong quality performance, it will be less effective. The same holds true for human resource development, strategic management, and other areas of infrastructure. A lack of understanding of this linked and mutually affecting relationship is one of the major reasons why some firms end up with an ineffective system after adopting JIT.

Thus, the HPM view of JIT is as a comprehensive approach, involving both core JIT practices and supporting infrastructure practices. Plants that achieve the best JIT performance, in terms of such standard JIT measures as inventory turns and reduced cycle time, also have strong infrastructure practices. This comes from understanding both the effect that infrastructure has on JIT performance and the effect that JIT strength has on pushing infrastructure practices to improve.

Evidence for these relationships was provided by analysis of the plants in our data set. We examined the relationship between groups of infrastructure practices and JIT practices by breaking down the practices used in plants with high JIT performance and in those with low JIT performance, measured as the inventory turnover ratio. We found that the level of JIT practices used between the two groups does not differ substantially. However, when the combination of JIT practices and infrastructure practices is examined, there is a difference in JIT performance. The plants with the best JIT performance have superior quality management, human resource development, and strategic management practices. Thus, there is a levered linkage between the practices. The infrastructure practices support the core JIT practices, helping to solve problems in the process and to hasten flows through the process, allowing JIT practices such as setup-time reduction to function more effectively. In turn, the JIT practices encourage development of better quality management, strategic management, and human resource development practices.

SUMMARY

We have presented the HPM perspective of JIT. Although there are certainly benefits associated with the use of JIT practices alone, we have found that the plants with the best JIT performance know how to exploit the linkages between JIT practices and supporting infrastructure practices. We focused on three primary areas of infrastructure (quality management, human resource development, and strategic management) in order to explain the relationship between infrastructure practices and JIT performance. However, there are other important areas of infrastructure, including information systems and product development practices, that also offer potential synergies with JIT performance.

We also examined the role of JIT as a path to competitive performance. We found that there were some JIT paths that seemed to lead to better performance; however, we also found evidence that JIT is not the only path to achieving the types of performance we associated with JIT, including annual inventory turns and short cycle times. For example, the plants in the United Kingdom were competitive on annual inventory turns without strongly following the JIT path. Inventory turnover is the ratio of annual sales to the aggregate inventory level, so it is easy to see that this ratio can be improved by focusing on either increases to annual sales or decreases to aggregate inventory levels. Likewise, there are practices associated with cycle-time reduction that are not associated with JIT. Thus, one path to high performance in flow-related measures is JIT; however, there are other equally viable paths. It is important for each plant to assess which practices are most appropriate for its unique environment and culture.

The managerial implications of this are clear. A plant should start by developing its manufacturing strategy and its associated approach to JIT. Managers should assess the degree of infrastructure already in place before beginning JIT implementation. They should decide if JIT is the best path to improvement or whether further infrastructure development or other approaches should be taken. Once the manufacturing strategy is developed, it will guide the eventual implementation of the appropriate practices including core JIT techniques. This will provide a tailor-made approach to JIT implementation in each manufacturing plant.

References

Bonazzi, Giuseppe. "Coalitions in a Japanese Transplant in Italy." *International Executive,* 37, no. 4 (1995): 305–414.

Monden, Yaguhiro. *Toyota Production System: Practical Appraoch to Production Management.* Norcross, GA: Industrial Engineering and Management Press, 1983.

Monden, Y. "What Makes the Toyota Production System Really Tick?" *Industrial Engineering* (January 1981): 36–46.

Schonberger, R.J. *Japanese Manufacturing Techniques: Nine Hidden Lessons in Simplicity.* New York: Free Press, 1982.

Sugimore, Y., K. Kusunoki, F. Cho, and S. Uchikawa. "Toyota Production System and Kanban System—Materialization of Just-in-Time and Respect-for-Human System." *International Journal of Production Research* 15, no. 6 (1977): 553–564.

CHAPTER 9

QUALITY: FOUNDATION FOR HIGH PERFORMANCE MANUFACTURING

BARBARA B. FLYNN, E. JAMES FLYNN, and
ROGER G. SCHROEDER

Of all the practices that we have discussed as elements of high performance manufacturing (HPM), quality management is probably the best known. There have been literally thousands of articles written about what quality management is and what its critical practices are. In addition, there are prescriptive standards, such as ISO 9000, and award criteria, such as the Malcolm Baldrige National Quality Award *Criteria for Excellence,* that describe ideal quality management practices. Many organizations follow a specific approach, such as those based on the work of Crosby, Deming, or Juran.

However, what is less well known about quality is how it functions as a source of competitive advantage. Although we know that quality can be a potent source of competitive advantage, how does it relate to other competitive priorities? Is it possible for a plant to achieve top performance in quality and in other dimensions of competitive performance simultaneously? Or does top performance in quality come only at the expense of other dimensions of competitive performance? For example, are quality and cost always a tradeoff? What about quality and product development speed—is it necessary to take quality shortcuts in order to increase the speed of a new product to market?

At the heart of this discussion is a long-running debate about whether dimensions of competitive performance necessarily function as tradeoffs or whether there are synergies possible between certain dimensions of competitive performance, including quality. The tradeoffs perspective suggests that a plant that achieves quality leadership in its industry should not attempt to also become a leader in terms of cost, dependability, speed, or flexibility. In

163

fact, doing so could destroy the quality advantage the plant had already achieved because it is inherently impossible to achieve top performance on multiple dimensions of competitive performance simultaneously.

However, the well-known quality "guru" Phil Crosby suggested that "quality is free." He stated that it is much more costly to produce poor quality because the true cost of quality includes the often forgotten costs of inspection, the resulting rework, and scrap costs and the potential costs of loss of goodwill and warranty fulfillment. Although quality training, reengineering, and design for manufacturability can initially be costly, the benefit comes from the lowered cost of not having to operate the additional space, machines, material, and people required to repair defects and to deal with product failures in the field. Crosby's notion of "doing it right the first time" popularized the idea that high quality could be achieved at a relatively low cost.

The underlying philosophy of HPM supports this approach, based on the work of Schonberger, who extended Crosby's "quality is free" notion to other dimensions of competitive performance. For example, as quality improves, cycle time should be reduced through elimination of inspection operations and rework. The HPM perspective suggests not only that it is *possible* to achieve top performance on several dimensions of competitive performance simultaneously, but also that it could actually be *simpler* to pursue several dimensions of competitive performance simultaneously.

How could this be? We suggest that development of an infrastructure of HPM practices will make the achievement of top performance on several dimensions of competitive performance seem natural, rather than contradictory. Thus, there are compatibilities, or synergies, between some dimensions of competitive performance. Because quality management practices form an important part of the HPM infrastructure, a plant that excels in quality should be strongly positioned to compete on other dimensions of competitive performance as well.

In this chapter, we look at how quality relates to other dimensions of competitive performance. Is it really possible to achieve synergies between quality and other dimensions of competitive performance? If so, how common is this strategy? We also examine patterns of tradeoffs and compatibilities in the five countries we studied to determine whether there are differences in competitive performance strategies based on national culture. Finally, we look at what comprises the common infrastructure of HPM practices that supports the achievement of compatibilities between quality and other dimensions of competitive performance.

Tradeoffs versus Compatibilities

The Tradeoffs Perspective

The tradeoffs perspective has been the dominant way of operating in the manufacturing function for many years. It was first introduced in 1969 by Wickham Skinner, who described the relationship between product cost and other product attributes (including high quality, low inventory levels, a wide range of products, and short, reliable delivery times) as a tradeoff. He maintained that a tradeoff will occur if a plant attempts to pursue two goals, because the goals are inherently conflicting. We note that Skinner's definition of quality was not conformance to specifications or to customer requirements, but rather product attributes. Enhancing product attributes, or product features, without changing the technology usually increases costs. In other words, a Cadillac costs more than a Chevrolet.

Another perspective is presented by the British scholar Terry Hill. He advocates that a plant must align its operations function with the volume level targeted by the marketing function. Volume level should dictate the organizational infrastructure, including competitive priorities. For example, an operation designed to produce a narrow product range at a high volume must necessarily compete on cost, whereas a low-volume, more customized operation may compete on quality, product features, or capability but should not compete on cost. The best plants will be those that focus tightly on a particular volume level and its associated infrastructure. Thus, many dimensions of competitive performance are incompatible at a given volume level.

Michael Porter is perhaps the strongest advocate of the tradeoffs perspective. He identified two types of competitive advantage:

1. A *lower cost* advantage occurs when a plant designs, produces, and markets its products more efficiently than its competitors.
2. A *differentiation* competitive advantage exists when a plant can make its product unique from its competitors' products in some way, including quality, service, brand image, or distribution.

Porter stresses that a plant *must* make a choice about which competitive advantage it will pursue. Pursuing both cost and differentiation strategies is a "recipe for strategic mediocrity," resulting in a plant becoming "stuck in the middle."

The tradeoffs perspective can be seen in many decisions made in plant manufacturing functions. Statements such as the following are often heard in plants that subscribe to the tradeoffs perspective:

- "We haven't pursued quality management more aggressively because we just don't have the money to invest in it right now."
- "We can't shorten product development times any further without cutting corners and sacrificing quality."
- "We could improve delivery times a great deal if we build more inventory."

Such statements imply that an organization must pursue a single objective because these objectives are inherently contradictory. Until recently, this philosophy was widely embraced in manufacturing and supported by traditional economic analysis. However, the evidence from Japanese manufacturers in the late 1970s and the 1980s indicated otherwise. Despite the expectations of economic theorists, many Japanese plants were routinely producing extremely high-quality products at a very low cost. More recently, some Japanese plants have added short product development time to their list of competitive priorities.

The economists are not entirely wrong; rather it is a matter of assumptions. Economists assume that plants operate at the efficient frontier to achieve the best performance possible from a given set of resources. In using a static model, they also assume that technology is fixed and cannot be improved. As a result of these two assumptions, tradeoffs will occur. In reality, plants do not operate at the efficient frontier; they have slack in their operations. Also, technology is constantly changing and moving the frontier forward. Accordingly, plants can improve on multiple dimensions of performance. The static economic assumptions that have led to tradeoff theory are usually not met in actual practice.

The Compatibilities Perspective

The evidence from Japanese plants in the late 1970s and the 1980s led to the development of the world class manufacturing perspective. This perspective suggests that the ability to develop simultaneous competitive advantages is achieved through development of an infrastructure of practices focused on controlling processes, designing processes to consistently produce the product correctly, and continuous improvement of product and process management.

Spearheaded by the work of Richard Schonberger, the world class manufacturing perspective suggests that the tradeoff perspective is a myth, outdated by the adoption of world class practices, which stimulate solutions to quality problems, thus eliminating unnecessary inventory and reducing waste and processing. These results lead to cost compression and lead-time reduction. HPM builds on the world class manufacturing perspective.

How can compatibilities be possible, despite dire predictions of strategic mediocrity? The approach developed by Ferdows and DeMeyer in 1990 (see Figure 9.1) describes manufacturing competitive priorities as cumulative. Plants that manage their strategy by pursuing competitive priorities in a specific sequence will be more successful than those that do not. What does this mean? Specifically, quality provides the foundation for dependability, which provides the foundation for speed. Only after speed capabilities have been developed should cost be addressed. The reason for this is the infrastructure that is developed as each of these dimensions is pursued,

Figure 9.1
The Sandcone Model

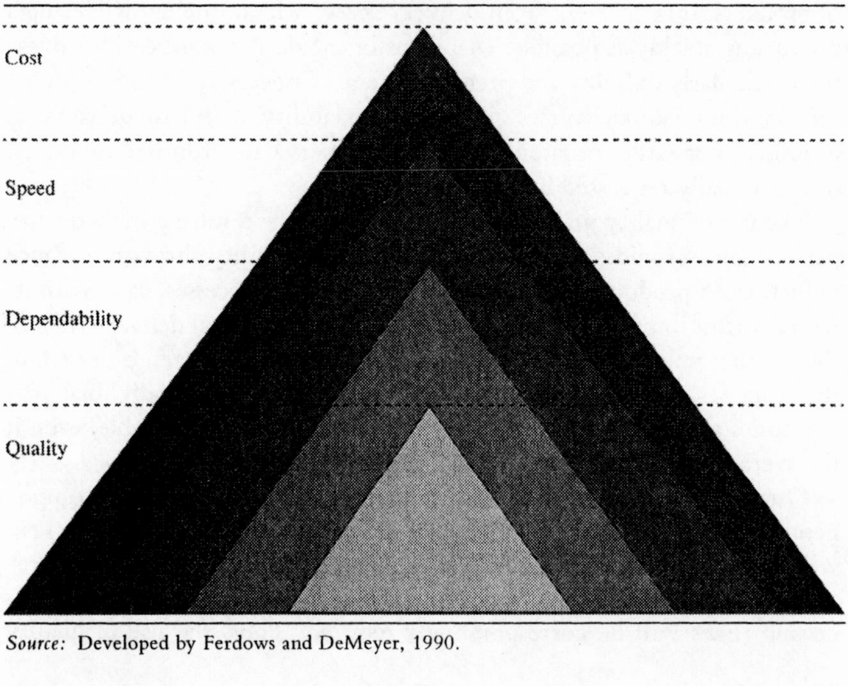

Cost

Speed

Dependability

Quality

Source: Developed by Ferdows and DeMeyer, 1990.

with quality management practices at its foundation. Thus, quality provides the base for long-term improvements in other dimensions of competitive performance.

The Ferdows-DeMeyer approach is referred to as the *sandcone model* of manufacturing performance, using the analogy of a child pouring sand through a funnel at the beach. As the cone that is formed under the funnel grows taller, it simultaneously grows wider. Thus, the foundation of the sandcone grows to support its height. Translated into manufacturing terms, the quality foundation must continue to expand as an organization broadens its horizons to pursue the additional competitive dimension of dependability. Speed should be pursued after dependability and quality; however, the foundation provided by quality and dependability should continue to be expanded as speed is added to the mix. The competitive dimension of cost should be pursued only by organizations that already have the infrastructure foundation provided by quality, dependability, and speed.

How do these specific relationships work? What allows them to function as compatibilities rather than as tradeoffs? They are each described in the following sections.

Quality and Dependability. *Dependability* refers to the on-time delivery of products. It is different from delivery *speed,* which implies delivering a product as quickly as possible. Many customers do not want early deliveries, particularly of bulky and perishable items. Conversely, even fewer customers want late deliveries. Thus, dependability refers to delivery as promised, not earlier or later. Ideally, the delivery date promised is also the date originally requested by the customer.

The use of quality management practices and the resulting high-quality performance provides the foundation for dependability through variance reduction. A product is produced by a sequence of processes, each with its own distribution of possible completion times. The actual delivery time of the product will be the sum of the times that the product spends in each of the processes, including waiting and transport times. When individual process times are more variable, delivery times will be more variable, even if the average processing time for each step is the same.

Quality management practices reduce the variability of the components of processing times through the use of approaches designed to help workers understand the causes of variability so they can subsequently reduce them. As the sources of variability are reduced, the variability of processing times will be correspondingly reduced. Thus, the use of quality management practices lays the foundation for improving performance of both quality and dependability.

Quality and Speed. The idea of speed as a dimension of competitive performance has received a great deal of attention in recent years, particularly as Japanese organizations have excelled at speedy product development and cycle-time reduction. Also described as "time-based competition," speed is heralded as the feature most market leaders have in common. *Speed* is defined as the total time required to design and deliver a product. Thus, speed includes two dimensions—(1) delivery time and (2) design time—both with potential synergies with quality.

Delivery speed deals with compression of the time of the production process. The use of quality management practices can help reduce delivery time through both inventory reduction and cycle-time reduction. As the inventory turnover ratio increases, there are more dollars in sales for every dollar in inventory; thus, products are getting out the door faster. Inventory turnover is improved by high-quality levels, through reduction of the need to maintain large amounts of safety stock inventory to compensate for the absence of a constant work flow. Quality management practices, such as design for manufacturability, also facilitate setup-time reduction, leading to reductions in cycle stock inventories. Pipeline inventories are reduced through the improved flows that result from the use of quality management practices.

Cycle times are reduced as the use of quality management practices reduces the number of items requiring inspection and rework. The use of certified suppliers and long-term supplier relationships based on quality criteria can reduce or eliminate preprocessing cycle-time delays for incoming inspection. Practices such as quality at the source, feedback, statistical process control, and effective product designs reduce delays for rework and process inspection of in-process and finished goods. Thus, the use of quality management practices provides the foundation for reduction of delivery speed.

The second dimension of speed is *new product speed,* or the speed with which new products are brought to market. Many of the practices associated with quality management are very similar to the practices suggested for fast product innovation. For example, quality management practices associated with customer focus shorten product development times through better incorporation of customer desires. Quality management practices associated with employee involvement form the foundation for team-based concurrent engineering, which reduces product development times by involving people from diverse functional specialties in the design process. Continuous improvement activities reduce product development times through incremental innovation, which hones in on customer needs, leading to less design rework.

Quality and Cost.   Many managers believe that quality management initiatives are costly and that cost cutting is done only at the expense of quality. In fact, traditional economics-based models have focused on specifying the optimal level of quality, based on the tradeoff between the costs of nonconformance and the costs of operating a quality management initiative. However, this thinking assumes that the process technology is fixed and cannot be changed. In fact, process technology is constantly being improved, thereby offering opportunities for both quality improvement and cost reduction. Often, managers only look at the cost of training, inspection, and other costs of implementing a quality management initiative when they estimate the cost of quality. However, the true cost of quality should also include the costs associated with avoiding poor quality, such as inspection, as well as the costs that occur as a result of producing poor quality, such as rework and warranty costs. Thus, quality costs include four types of costs:

1. *Prevention costs* are the costs associated with planning the quality system, process control development costs, information systems costs, training costs, and general management costs.
2. *Appraisal costs* include the cost of finding defective items, test and inspection costs, the costs of maintaining instruments, and process-control implementation costs.
3. *Internal failure costs* are the costs associated with defects that are found before the product leaves the plant, including scrap and rework costs, costs of corrective action, and downgrading costs.
4. *External failure costs* are the costs of defective products that reach the hands of the customers, including costs of customer complaints and returns, product recall costs, warranty claims costs, product liability costs, and goodwill costs.

In a plant that uses a traditional approach to quality management, the highest costs are appraisal costs, internal failure costs, and external failure costs (see Figure 9.2). By contrast, plants that have high-quality performance place a much greater emphasis on prevention costs; and their appraisal,

Figure 9.2
Costs of Quality

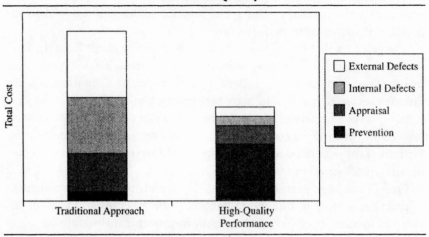

internal failure, and external failure costs will be much lower. The end result is that the total cost of quality will be lower in a plant with high-quality performance, even though the plant may spend considerably more on prevention costs as compared to plants with less emphasis on quality. Thus, there may be the appearance of a tradeoff when prevention costs increase as quality management is implemented. However, plants with high-quality performance will have lower total costs of quality because costs in appraisal, internal failure, and external failures costs are considerably less.

TRADEOFFS AND COMPATIBILITIES ACROSS COUNTRIES

High- and Low-Performing Plants

What did we find out about quality as we studied the plants in our database? We began by dividing the sample into plants with high-quality performance and plants with low-quality performance. We looked at a total of nine quality management practices. We found that there was a statistical difference between the high- and low-performing plants on six of the nine practices: (1) small-group problem solving, (2) feedback, (3) process control, (4) supplier quality involvement, (5) top management leadership for quality, and (6) rewards for quality. Figure 9.3 shows that for each of the six practices, the high-performing plants had a higher score. Thus, the high-performing plants pursued this set of practices more vigorously than the plants with low-quality performance.

The practices that differentiate the high-performing plants from the low-performing plants cover a comprehensive range of areas, including employee involvement, process focus, leadership, and suppliers. This illustrates the linkages between practices in high-performing plants. What is interesting is that this set of practices does not include the practices related to a plant's relationship with its customers (customer involvement and customer satisfaction). This occurs because both high- and low-performing plants emphasize customer relationships.

One of the high-performing plants that we visited used a broad range of practices leading to high performance. In the beginning, the plant decided to involve its employees in quality improvement through the use of employee-led quality teams. Top management instigated this program and remained visible in rewarding and encouraging the improvement teams. Later, plant management started employee training in statistical techniques and instituted statistical process control. This was followed by gaining more input from customers and then from suppliers. This plant followed a typical path that many high-performing plants have used to improve their quality.

Figure 9.3
High Performance Quality Practices

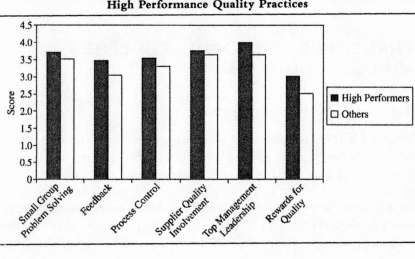

We also looked at quality management practices by country. We found that there were seven practices that were significantly different across countries. These are shown in Figure 9.4 ((1) small-group problem solving, (2) customer involvement, (3) customer satisfaction, (4) feedback, (5) process control, (6) top management leadership for quality, and (7) rewards for quality).

Not surprisingly, the Japanese plants were the highest on every practice except two. What was surprising was that Japanese plants were not the highest on the two practices related to relationship with customers (customer involvement and customer satisfaction). In fact, it was the lowest of the five countries on each, despite the fact that Japan had the largest number of plants with high-quality performance. What does this mean? It appears that Japanese plants have achieved very high quality with less emphasis on customer relationships than other countries. This is not to say that customer relationships aren't important in Japan; rather, they may be so tightly integrated in Japanese plants that they require less attention to maintain than they do in other countries where quality management efforts are less well developed. The Japanese plants were particularly strong in top management leadership for quality, rewards for quality, and feedback.

One plant that we visited in Japan demonstrated its tremendous commitment to quality. The workers had been very actively involved in quality circles since the late 1960s. There was evidence of statistical quality control charts and feedback to employees throughout the plant. When we talked with top management, they expressed their commitment to quality as the number-one goal. This plant exhibited a comprehensive approach to quality improvement in all parts of the plant.

Continuing to examine Figure 9.4, we find that the U.S. and the German plants are well positioned to compete with the Japanese. They lead in the areas of customer satisfaction and customer relationships. But, the U.S. and the German plants trail the Japanese plants in many areas. This indicates that high-performing plants do not need to be good in everything. They can choose their own path to high performance.

Compatibilities and Tradeoffs

After reviewing performance, we addressed compatibilities and tradeoffs. We looked at a total of seven different measures of the four dimensions of competitive performance as follows:

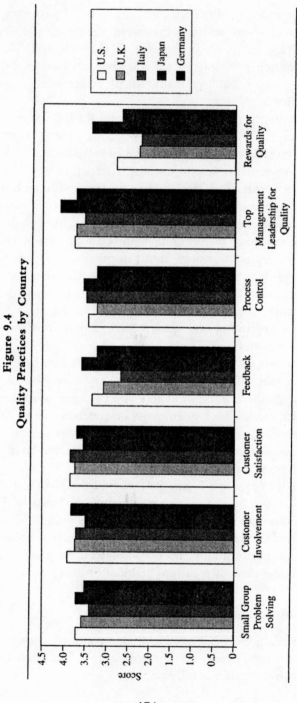

Figure 9.4
Quality Practices by Country

1. *Quality*
   - Internal quality: percent of internal scrap and rework.
   - External quality: percent of defective products returned.
2. *Cost*
   - Total manufacturing cost as a percent of sales.
3. *Dependability*
   - Percent of orders shipped on time.
4. *Speed*
   - Cycle time: time from receipt of raw materials through delivery of the product to the customer.
   - Inventory turns: ratio of cost of sales to aggregate inventory.
   - New product speed: number of months to introduce a new product.

To determine tradeoffs and compatibilities, we ranked the performance of each plant in its industry on each of the measures of competitive performance. This ranked list of plants in their industries was then divided into four equal parts, or quartiles. A plant's quartile ranking on a given measure of competitive performance showed its position within the industry on that measure. To our surprise 78 percent of the plants achieved performance in the top quartile of their industry on at least one of these dimensions of competitive performance (Table 9.1). This illustrates that there are very few "bad" plants in our database; only 22 percent were not in the top quartile in their industry on at least one dimension of competitive performance. On the other hand, very few plants were able to dominate on three or four measures of performance. Only one plant out of 164 dominated on all four performance measures, and only 14 plants were able to dominate on three out of four measures. This indicates that plants can

Table 9.1
Number of Plants with High
Performance on Dimensions of
Competitive Performance

Number of Plants	Number of Dimensions
36	0
63	1
50	2
14	3
1	4

often find a niche or a way to compete. It is not easy or likely for a competitor to dominate on more than two dimensions of performance.

We compared quartile rankings between pairs of measures of competitive performance in order to determine the presence of tradeoffs or compatibilities. If a plant was in the top quartile on one dimension of competitive performance and in the bottom quartile on another dimension of competitive performance, we considered that to be a tradeoff. In contrast, if a plant was in the top quartile on two dimensions of competitive performance, we considered that to be a compatibility between those two dimensions of competitive performance. We considered all other possible conditions to be neither a tradeoff nor a compatibility.

Compatibilities did exist between quality and other dimensions of competitive performance. Of the 164 plants in the sample, 41 (one-fourth of 164) fell in the top quartile on any given dimension of competitive performance, by definition. Thus, there was the possibility of 205 compatibilities (41 × 5 dimensions of competitive performance besides quality). We found that there was a compatibility between quality and another dimension of competitive performance in 16 percent of the possible instances.

Tradeoffs between quality and other dimensions of competitive performance also occurred; however, the dominant situation was that quality and other dimensions of competitive performance represented neither tradeoffs nor compatibilities. Thus, our findings did not support the idea that quality is always traded off against cost, speed, and so on, but they didn't indicate that compatibilities were the norm either. What our analysis does show is that some plants have combined top performance in quality with top performance on other dimensions of competitive performance. These are the high-performing plants. It isn't necessarily easy to combine quality with other dimensions of competitive performance, but it is the goal for which high-performing manufacturers strive.

What this illustrates for managers is the need to develop a strategy and a niche in order to compete. It is not a good idea to adopt every practice that comes along or to only seek "best practice" as the way to go. Each plant should develop the unique set of practices and approaches to quality that gives it a competitive edge with respect to its customers and its particular situation. Determining that competitive niche and gaining agreement of all to pursue it is the challenge to management.

We also looked at whether quality was pursued simultaneously with other dimensions of competitive performance in a similar fashion across countries. Plants in Japan had the greatest percent of compatibilities between *internal quality* and other dimensions of competitive performance. This was followed by plants in Italy and the United States, with a lower emphasis on pursuing both internal quality and other dimensions of competitive performance. Plants in the United Kingdom and Germany were the lowest, with very little emphasis on compatibilities between internal quality and other dimensions of competitive performance. Japanese plants use strength in internal quality as a platform for achieving strength in other dimensions of competitive performance. They focus their efforts on minimization of scrap and rework through process control and improvement. This is illustrated by the reports of manufacturing excellence in Japan in the management and popular literature for the past few decades.

In contrast, Italy and the United States had the largest percent of compatibilities between *external quality* and the other dimensions of competitive performance. This was followed by a somewhat lessened emphasis on compatabilities between external quality and other dimensions of competitive performance by Germany, the United Kingdom, and Japan. With the exception of Japan, all the countries had a substantially greater emphasis on compatibilities between external quality and other dimensions of competitive performance than on compatibilities based on strength in internal quality. Plants in countries other than Japan use a strategy of building strength in other dimensions of competitive performance based on their excellence in external quality, indicating a somewhat stronger focus on customers relative to internal operations as compared with Japanese plants. This is consistent with the lower emphasis on customers by Japanese plants that we noted earlier.

We have noticed that plants in the United States, for example, have a very strong focus on customer satisfaction. Many of them have programs aimed at "customer driven quality," which reflects their intent not only to relate to the customer but also to be driven by the customer. This approach is very consistent with our data. As the same time, plants should be determined to put a strong infrastructure behind their customer focus, as the Japanese plants have demonstrated.

Strategies for Competitive Performance

Which particular dimensions of competitive performance are pursued simultaneously with quality in each of the countries? We found that there were two types of strategies: (1) selective strategy and (2) comprehensive strategy. Plants in both Italy and Germany followed a *selective strategy,* emphasizing compatibilities between quality and a few other dimensions (see Table 9.2). In both countries, quality/cost and quality/new product speed compatabilities were emphasized. Plants in the United Kingdom followed a similar selective strategy, however, at a lower level, emphasizing quality/new product speed and quality/on-time delivery compatibilities. In contrast, plants in Japan and the United States followed a *comprehensive strategy*, appearing to strive for compatibilities between quality and each of the other dimensions of competitive performance.

The concept of selective and comprehensive strategies was first introduced in Chapter 2. We reinforce this concept here with our observations about quality and note that plants in different countries adopt different paths to HPM. We wondered whether plants that followed different compatibility strategies did so by emphasizing different practices. We statistically determined the practices associated with various compatibility strategies and have summarized them in Table 9.3. There are a number of practices related to quality/dependability, including practices such as human resource (HR) management, information systems, strategic management, quality management, and JIT. This compatibility is built on across-the-board excellence, so it may be more difficult to achieve than other compatibilities.

There were fewer practices associated with differences between the groups of plants for the relationship between quality and other dimensions of competitive performance. For example, the only practices that differentiated between the plants for quality/new-product-development speed were practices associated with quality management, including pride in work, rewards for quality, feedback, process control, and supplier quality involvement. Thus, there is an infrastructure of practices that supports both quality and product-development speed. Quality management provides the foundation for

Table 9.2
Pairs of Dimensions Compatible with Quality by Country

	Germany	Italy	Japan	United Kingdom	United States
	Percentage of Pairs of Dimensions Compatible with Quality				
Cost	21.8%	21.1%	15.6%	3.8%	12.9%
Cycle time	9.7	2.4	10.4	3.8	10.3
Inventory turnover	0.0	9.4	12.2	7.6	10.3
New product development speed	14.5	21.1	8.7	11.4	15.5
On-time delivery	4.8	4.7	12.2	11.4	12.9

Table 9.3
Relationship between Practices and Compatibilities with Quality

Discriminating Variables (Practices)

Quality/Dependability Compatibility

Human resource management	Manufacturing/HR fit
	Centralization of authority
	Shop floor contact
	Compensation for breadth of skill
	Small-group problem solving
	Documentation of shop floor procedures
	Coordination of decision making
	Multifunctional employees
	Recruiting and selection
Information systems	Internal quality information
	Dynamic performance measures
	Accounting information systems
Strategic management	Communication of manufacturing strategy
	Formal strategic planning
	Anticipation of new technologies
Quality management	Top management support for quality
	Rewards for quality
	Feedback
	Process control
	Supplier quality involvement
JIT	Setup-time reduction

Quality/New Product Development Speed Compatibility

Quality management	Pride in work
	Rewards for quality
	Feedback
	Process control
	Supplier quality involvement
Human resource management	Manufacturing/HR fit
	Employee suggestions—implementation and feedback
	Coordination of decision making
	Documentation of shop floor procedures
	Supervisory interaction facilitation

Quality/Delivery Speed Compatibility

Information systems	Dynamic performance measures
	Internal quality information

Quality/Cost Compatibility

Human resource management	Manufacturing/HR fit
	Multifunctional employees
	Compensation for breadth of skill

pursuing new-product speed as an additional source of competitive advantage (in addition to quality). This compatibility was emphasized in four out of the five countries we studied, perhaps because it is relatively easy for plants to achieve high performance in new-product-development speed when they already have achieved high performance in quality management.

The practices associated with differences between the groups on quality/delivery speed compatibility were focused on human resource management and information systems. Thus, plants that excelled on both quality and delivery speed had a very different infrastructure from those that excelled on quality/dependability or quality/new-product-development speed. This provides an explanation for why the presence of compatible dimensions of competitive performance was relatively rare. Plants need to develop very different infrastructures in order to excel on different simultaneous dimensions of competitive performance. Although differences in JIT practices were expected to contribute to differences in quality/delivery speed compatibilities, this was not the case. The practices associated with differences between the groups with compatibilities, with tradeoffs, or with neither condition between quality and cost were all practices related to human resource management.

These findings point to a sandcone of practices to support compatible dimensions of competitive performance. It suggests that plants that pursue quality/delivery speed after quality/dependability are better positioned to ultimately achieve quality/cost compatibility than those that focus instead on quality/new-product compatibility because they will have developed the strength in human resource practices needed to pursue quality/cost compatibility. This is the route followed by the Japanese plants, which had the largest number of compatibilities. The plants in the other countries emphasized the quality/new-product-development speed route, limiting their potential for achieving further compatibilities through failure to continue to develop HR compatibilities.

One of our most interesting findings is that compatibilities between quality and other dimensions of competitive performance are common, not the rare and unusual occurrence suggested by the tradeoffs school of thought. A related conclusion is that plants should develop different infrastructures to support different compatibilities with quality. Perhaps the most intriguing finding is the difference in the extent of compatibilities between dimensions of competitive performance and quality. With the exception of Japan, compatibilities with other dimensions of competitive performance are more common with external quality than with internal quality.

We conclude with a word of advice for managers. First, identify your strategy or niche for competing. This may require excellence (e.g., top quartile) in more than one dimension of manufacturing performance. But be careful not to attempt to excel at everything; very few plants can excel on all dimensions. Select those dimensions that are truly important for your strategy. Next, managers should develop the infrastructure that will support the competitive dimensions selected. This will require management to ask, "Which practices are the most appropriate for our situation and strategy?" As a result, a unique and defined set of practices will emerge for implementation. Whatever practices are selected should then be carefully integrated and linked together with existing and emerging practices.

We also recommend that the sandcone model be used as a way to proceed. The base should be built first around quality, then dependability, speed, and finally cost. This means that a plant seeking low cost as its primary dimension of competition cannot get there in one step. The plant must first build the base in quality infrastructure and proceed along the sandcone model to cost. The order of implementation is important to achieve certain performance dimensions.

Our final word of advice is to implement the chosen strategy and path with vigor and determination. There is not only one way to compete, but whatever way is chosen must be vigorously pursued.

SELECTED READINGS

Crosby, Phillip. *Quality Is Free.* New York: McGraw-Hill, 1979.

Deming, W. Edwards. *Out of the Crisis.* Cambridge, MA: Massachusetts Institute of Technology, 1986.

Ferdows, Kasra, and Arnoud De Meyer. "Lasting Improvements in Manufacturing Performance: In Search of a New Theory." *Journal of Operations Management* 9, no. 2 (April 1990): 168–184.

Flynn, Barbara B., Roger G. Schroeder, and Sadao Sakakibara. "A Framework for Quality Management Research and an Associated Measurement Instrument." *Journal of Operations Management* 11, no. 4 (March 1994).

————. "Determinants of Quality Performance in High and Low Quality Plants." *Quality Management Journal* (August 1994).

————. "The Impact of Quality Management Practices on Performance and Competitive Advantage." *Decision Sciences* 26, no. 5 (1995).

———. "The Relationship between Quality Management Practices and Performance: Synthesis of Findings from the World Class Manufacturing Project." In *Advances in the Management of Organizational Quality,* ed. Donald Fedor and Soumen Gosh. JAI Press, 1996.

Hill, Terry. *Manufacturing Strategy: Text and Cases,* 2d ed. Burr Ridge, IL: Irwin, 1994.

Juran, Joseph M. *Juran on Leadership for Quality: An Executive Handbook.* New York: Free Press, 1989.

Porter, Michael. *Competitive Strategy: Techniques for Analyzing Industries and Competitors.* New York: Free Press, 1980.

Schonberger, Richard. *World Class Manufacturing.* New York: Free Press, 1986.

Skinner, Wickham. "Manufacturing: Missing Link in Corporate Strategy." *Harvard Business Review* (May–June 1969): 136–145.

PART III

HIGH PERFORMANCE MANUFACTURING BY COUNTRY

CHAPTER 10

U.S. MANUFACTURING RENEWAL

ROGER G. SCHROEDER and BARBARA B. FLYNN

The United States remains strong in manufacturing, despite some turbulent years beginning in the 1970s. In this regard, the U.S. automobile industry is a good example of changing global markets and the resultant impact on the U.S. auto industry. In 1955, U.S. automobile companies held 99 percent of the U.S. auto market, with only a few imported European luxury cars. By 1970, however, this situation had changed drastically, with 15 percent of the U.S. market being held by imports (mostly Japanese); and by 1985, imports held 25 percent of the U.S. auto market (Womack, Jones, and Roos, 1990). It was not until the early 1990s, after losing more than a quarter of the U.S. auto market to Japanese car makers, that the flow of imports was finally stemmed. In the meantime, over a thousand Japanese plants had been built in the United States, not only in the automobile industry but also in electronics, machinery, and other industries aimed at solidifying the Japanese position in the U.S. market.

This illustrates three points common to changes in U.S. manufacturers during the period of 1970 to 1990.

1. The loss of U.S. market share in the 1970s and 1980s to foreign competition, mostly to Japanese and other Asian competitors.
2. The location of foreign plants on U.S. soil to directly attack the U.S. market.
3. The globalization of industries (auto, electronics, machinery, etc.) that were formerly dominated by U.S. companies.

In the 1990s, however, U.S. companies were able to respond to these challenges, and the United States has restored its manufacturing industry to world class competitive standards today.

185

Beginning in the 1970s and 1980s, U.S. manufacturing companies slowly realized that they were behind and that they needed to do something different to compete. For example, the total quality management (TQM) movement in the United States received a huge boost in public awareness in 1980 by the NBC white paper with W. Edwards Deming entitled, "If Japan Can, Why Can't We?" This telecast demonstrated to U.S. companies why they were behind Japanese companies and what would be required to raise quality levels and quality processes to world class levels. Many U.S. firms were shocked to learn that their products and processes did not measure up to world class standards. The Malcolm Baldrige National Quality Award, instituted in 1987, further highlighted the importance of quality and provided thousands of U.S. firms with a "blueprint" for quality improvement. The first Baldrige winner, Motorola, demonstrated the determination of U.S. companies to come back against vigorous Japanese competition and to once again compete at world class levels.

In the early 1980s, another Japanese innovation came to U.S. shores. The Kawasaki plant in Lincoln, Nebraska, attracted a great deal of attention for its implementation of just-in-time (JIT) practices along with some General Electric (GE) plants in 1980. This was followed by implementation of JIT in a number of U.S. factories throughout the 1980s and 1990s. Schonberger's first book, *Japanese Manufacturing Techniques*, was published in 1982, and his second book, *World Class Manufacturing*, was published in 1986, thereby accelerating the movement toward JIT and TQM in the United States. The latter book contained an "Honor Roll" of U.S. plants that had already implemented world class manufacturing practices, achieving 5-fold, 10-fold, or 20-fold reductions in manufacturing lead time (the 5–10–20s). Some of these plants were included in the U.S. world class sample for this study. At the same time, a group of researchers at the Massachusetts Institute of Technology (MIT) were developing the book *The Machine That Changed the World*. It chronicled the path of the Japanese auto industry and the response of the U.S. auto industry during the 1980s.

TQM and JIT are just two examples of the high performance manufacturing (HPM) practices implemented by U.S. companies to achieve a comeback in manufacturing performance. The result of the aggressive use of these practices, combined with favorable macroeconomic conditions, is dramatic. Figure 10.1 shows the growth in U.S. manufacturing productivity (output per hour) over the period from 1977 to 1998. Although productivity growth was almost zero during the late 1970s, it picked up in the 1980s, only to

Figure 10.1
Productivity Trends for U.S. Manufacturing

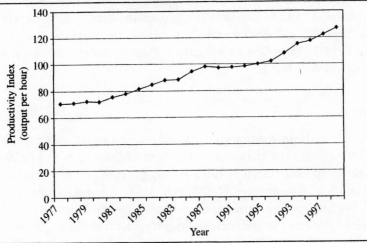

Source: Bureau of Labor Statistics, U.S. Government (http://stats.bls.gov), 2000.

slow again from 1988 to 1992. This five-year period (1988 through 1992) was followed by a return to normal levels. Since 1993, U.S. manufacturing productivity has been growing at a very healthy average rate of 4.1 percent per year, thanks to the resurgence of U.S. manufacturing and a very healthy U.S. economy.

At the same time, the very serious recession in Japan has taken the shine off the Japanese manufacturing miracle. During the 1990s the playing field in international manufacturing competition has been leveled, with U.S. firms able to compete with their Japanese and European rivals once again. Manufacturing is on the rebound in the United States.

A cynic might ask, "If manufacturing is so strong in the United States, why is there such a large trade imbalance in manufactured goods?" To be sure, manufacturing imports exceeded exports from the United States by an increasing amount in the later 1990s. To reverse this trend more on-shore manufacturing is needed in the United States. International financial arrangements may need to change for this to occur, along with further improvements in U.S. manufacturing productivity. We have some very good manufacturing in the United States. We just don't have enough of it to reduce the international trade deficit.

A TALE OF TWO MANUFACTURING COMPANIES IN THE UNITED STATES

The rebound in U.S. manufacturing, unfortunately, does not apply to all companies. We will describe two U.S. plants, one that has made the transition to HPM manufacturing practices (Plant A) and one that has not made the transition (Plant B).

Plant A

Plant A is in the machinery industry and makes a variety of equipment and machined parts. The plant was first built in a Midwestern city in 1968 and subsequently modernized in 1995. It engages in mostly repetitive production and has 200 employees (140 hourly and 60 salaried). Plant A has very high performance in various manufacturing measures, including low-cost on-time delivery, fast delivery, and quality, and is in the top 20 percent of all plants in the world in our sample. For example, the cycle time in Plant A is 17 days, 95 percent of the orders are filled on time, and quality is measured in parts per million defects, with zero returns for warranty or repair.

How did this plant achieve these impressive results? Was it sheer luck, management effort, the workforce, macroeconomic conditions, or what? We argue that its performance can be traced to the HPM practices and programs that have been implemented by management at this plant. For example, in the quality area, the plant has been very aggressive by certifying 100 percent of its suppliers. It has instituted statistical process control and the Plan-Do-Check-Act (PDCA) cycle for problem solving. This is being done not only by the engineers, but also by the workers, who are very active in solving quality problems in small groups.

Plant A has also been aggressive in JIT implementation. It has formed partnerships with customers and suppliers. It concentrates on reducing setup times and lot sizes through studying setups to reduce unnecessary elements, shifting internal setup activities to external setup activities, practicing setups, and using the reductions in setup time to justify smaller lot sizes. The plant uses a pull system to coordinate production, both within the plant and with suppliers and customers. As a result, Plant A has been able to drastically reduce inventories and cycle times.

In the human resources area, Plant A has been very successful at instilling pride in its workforce and building commitment that goes both ways between management and employees. Supervisors encourage teamwork, and they have been extremely good at fostering coordinated decision making and interfunctional cooperation. As a result, the climate for employees and for continuous improvement is very positive in this plant.

Plant A is strategically directed. It has done an excellent job of linking manufacturing to its business strategy. Its scores on a scale reflecting this linkage are among the highest of any plants that we have studied in the world. And it has done an excellent job of communicating its strategy to all its employees, as reported by the employees themselves. Finally, formal strategic planning is done on a regular basis in the plant and involves a significant portion of the employees.

Technology management and new product development is typified by a high level of interfunctional cooperation. But technology is not only planned across functions, it is also effectively implemented at the process level. Products are designed to be simple and producible from a manufacturing point of view.

Plant A sounds almost too good to be true. But it is an actual U.S. plant in our database that has implemented many of the practices that are described and measured in this book. It is a living example of how a plant can be managed using practices that are the envy of any plant in the world. Unfortunately, the same is not true of Plant B.

Plant B

Plant B is a U.S. plant that is on the opposite end of the spectrum. It makes products similar to those made in Plant A in the machinery industry, but it has a different approach to management. The plant is somewhat larger in size, with 395 employees (135 workers and 260 overhead). The performance of Plant B is in the bottom 20 percent of all plants in our database. It only fills 85 percent of its orders on time, inventory turnover is three times a year, and only 70 percent of the products are shipped without rework. The plant is still barely profitable, but it may not be for long, unless things change.

In the quality area, Plant B attempts to inspect quality into the product. It has 25 percent internal scrap and rework, and 15 percent of the products are returned from customers as defective. Clearly, Plant B is managing quality the wrong way. Table 10.1 shows the comparison between Plant A and Plant B. In terms of quality practices, Plant B is behind in every area of practice, including customer involvement, customer satisfaction, quality in new products, statistical process control, and small-group problem solving. The most striking difference, however, is in the area of top management leadership for quality improvement, where plant B has a leadership score of

Table 10.1
Comparison of Plant A and Plant B on Practices

	Plant A	Plant B	Difference (%)
Quality Management			
Customer involvement	4.31	3.67	17.4
Customer satisfaction	4.17	3.31	26.0
Quality in new products	4.11	3.32	23.8
Top management leadership for quality	4.38	2.86	53.1
Statistical process control	3.48	2.68	29.9
Continuous improvement	4.46	3.79	17.7
Small-group problem solving	4.35	2.95	47.5
Just-in-Time			
Partnership with suppliers	3.68	3.11	18.3
Daily run schedule adherence	3.09	2.86	8.0
Kanban	3.47	2.69	29.0
Small lot sizes	4.34	3.37	28.8
JIT link with customers	3.95	3.24	21.9
Setup-time reduction	3.62	2.71	33.6
Human Resources			
Commitment to the organization	4.26	3.57	19.3
Pride in work	4.47	4.28	4.4
Teamwork encouraged by supervisors	4.19	3.44	21.8
Multifunction employees	3.92	3.18	23.3
Employee suggestion implementation	4.23	3.50	20.9
Coordination of decision making	4.10	3.19	28.5
Manufacturing Strategy			
Formal strategic planning	3.67	3.22	14.0
Communication of strategy	4.13	3.06	35.0
Functional integration	3.75	2.33	60.9
Manufacturing strategy strength	3.73	2.60	43.5
Manufacturing-business strategy linkage	4.07	2.87	41.8
Technology Management			
Interfunctional design efforts	4.25	2.44	74.2
Effective process implementation	3.80	2.53	50.2
Product design simplicity	3.64	2.81	29.5

Source: HPM data.

2.86 on a 5-point scale; Plant A scores 4.38, a whopping 53 percent difference. Until top management takes responsibility and leadership for quality, the other quality practices will not improve in Plant B.

In terms of JIT, Plant B is also behind Plant A in every practice shown in Table 10.1. It has not formed partnerships with its suppliers or instituted setup-time reduction or Kanban to the same extent as Plant A. Plant B lags Plant A by from 8 percent to 34 percent in JIT practice areas.

As we look through human resources (HR), manufacturing strategy, and technology management in Table 10.1, we see a pattern emerging. Plant B is very poor in the areas of functional integration, effective process implementation, and interfunctional design efforts. In other words, it doesn't effectively communicate across functional boundaries and between management and employees. Aside from a lack of leadership, communication is sorely lacking in Plant B. As a result, very few of the practices that should be implemented are actually working. In fact Plant B isn't behind Plant A in only a few practice areas, it is behind in every practice area measured. Plant B actually started a TQM program in 1994, and it started employee involvement in 1991; but these programs have not taken root.

Plant A versus Plant B

Are Plant A and Plant B real aberrations? Did we select plants at the very extremes in order to make a point? Actually, they were selected because they are typical of high- and low-performing plants; we could have highlighted many other "Plant As" and "Plant Bs" from our data set.

Plant A clearly demonstrates the linkages between practice areas. It doesn't just excel in one area, it excels across the board. It is a plant that is constantly learning and applying its skills to new areas. Its success breeds success in a never-ending cycle of mutual reinforcement. Thus, levered linkages are very evident in Plant A. A levered linkage structure is typical of high-performing plants.

What about Plant B, then? It has pursued many of the same practices as Plant A. In fact, it even began some of these initiatives before Plant A. However, it never did the many things necessary to make sure that its new practices flourished. Thus, new programs and practices are viewed by the employees of Plant B as the "flavor of the month," without ever receiving their full commitment. New practices, poorly implemented, only divert management and employee focus away from core practices, leading to a downward spiral of performance. Thus, Plant B experiences "trapped linkages," where low performance in one area actually drags down performance in other areas. Recovering from this downward spiral can be extremely difficult. *Implementation* is the key difference between Plant A and Plant B, not actual practices employed or when they were first implemented.

We see from these examples that HPM manufacturing practices are being strongly implemented in the United States, as demonstrated by Plant A. Yet, there are still too many plants, like Plant B, that have only implemented these practices halfheartedly or not at all. In many cases the leadership, communication, and linkage among practices has been lacking.

COMPARISON OF U.S. PLANTS TO THE REST OF THE WORLD

But how do U.S. plants compare, in aggregate, to plants in other countries? In other words, does the United States have a preponderance of plants like Plant A when compared to other countries, or are U.S. plants more like plant B?

To answer this question, we have compared many of the scales in our study between plants in the United States, Germany, Italy, Japan, and the United Kingdom. In Table 10.2 we show the results by ranking the average U.S. plant scores from high to low. This results in three groups of scales, as follows.

1. Customers, Quality, and HR

In this group, the U.S. plants exceed all other countries on most scales. For example, on the customer involvement and customer satisfaction scales, the average U.S. plant does much better than the average Japanese plant. This comes as a surprise because Japanese plants are usually thought to be more customer oriented than U.S. plants. We believe this paradox can be explained by the impressions about the Japanese that are rooted in the 1970s and 1980s, while more recently U.S. plants are doing a much better job of serving their customers. Over time the most striking change in manufacturing in the United States has been "paying attention to the customer."

Another surprise is that the U.S. plants have a much higher score in top management leadership for quality than the Japanese plants (3.8 versus 2.5). Again, we feel that the U.S. top management has been stressing quality to a degree never before seen in the United States. In the meantime, Japanese plants have already achieved very high levels of quality practices, so that

Table 10.2
Ranking of U.S. Average Scale Scores from High to Low

	United States	Germany	Italy	Japan	United Kingdom
Customer involvement	**3.91**	3.83	3.71	2.87	3.75
Customer satisfaction	**3.86**	3.69	3.85	3.25	3.74
Multifunctional employees	**3.80**	3.63	3.33	3.58	3.61
Top management leadership for quality	**3.79**	3.73	3.53	2.48	3.71
Small-group problem solving	**3.72**	3.52	3.41	3.67	3.58
Cleanliness and organization	3.70	3.79	3.75	3.72	3.57
Benefits of information systems	**3.69**	3.51	3.49	3.29	3.64
External information—supplier quality control	3.66	3.67	3.72	3.69	3.64
Employee suggestions	**3.65**	3.50	3.32	3.62	3.39
Teamwork encouraged by supervisors	**3.64**	3.55	3.10	3.56	3.55
Equipment layout	3.62	3.58	3.60	3.67	3.51
Supplier quality involvement	3.62	3.68	3.64	3.70	3.69
Daily schedule adherence	3.61	3.57	3.73	3.51	3.31
Manufacturing-business strategy linkage	3.61	3.95	3.57	3.27	3.73
Commitment	**3.57**	3.35	3.48	3.37	3.33
Formal strategic planning	3.55	3.67	3.09	3.84	3.46
Communication of manufacturing strategy	3.50	3.65	2.94	3.68	3.38
Task-related training	3.48	3.24	3.25	3.73	3.26
Anticipation of new technologies	3.47	3.68	3.38	3.51	3.33
Dynamic performance measures	3.47	3.51	3.39	3.58	3.38
Product design simplicity	3.44	3.51	3.51	3.28	3.43
Documentation of shop floor procedures	3.44	3.53	3.39	3.01	3.41
Process control	3.43	3.24	3.47	3.60	3.23
Manufacturing strategy strength	3.43	3.64	3.48	3.55	3.42
Interfunctional design efforts	3.40	3.36	3.13	3.62	3.21
Internal quality information	3.37	3.60	3.29	3.76	3.56
Feedback	3.35	3.24	2.68	3.96	3.07
Coordination of decision making	3.35	3.60	3.27	3.83	3.27
Manufacturing/human resources fit	3.32	3.40	3.32	3.53	3.05
Shop floor contact	3.26	3.29	3.57	3.74	3.39
Recruiting and selection	3.25	3.20	3.14	3.45	3.08
MRP adaptation to JIT	3.25	2.94	3.03	3.51	3.00
Just-in-time delivery by suppliers	3.19	3.02	2.96	3.56	2.95
Coordination with corporation	3.12	3.72	3.73	3.17	2.80
Setup-time reduction	3.03	3.08	3.02	3.46	2.97
Maintenance	2.93	2.76	2.58	3.30	2.52
Accounting	2.90	3.24	3.21	3.76	2.65
Rewards for quality	2.79	2.64	2.21	4.03	2.24
Centralization of authority	2.68	2.45	3.19	3.50	2.83
Compensation for breadth of skill	2.59	2.86	2.74	3.68	2.54

Source: HPM data.

Note: Numbers shown in boldface type indicate rows in the table where the U.S. plants on average are higher than plants from all other countries.

quality is ingrained in the Japanese plant culture and no longer requires the same level of top management attention.

The third area where U.S. plants have an advantage is in small-group problem solving. In this case, however, the United States does not exceed the Japanese implementation level by very much, but it does have a substantial lead on Italian, German, and U.K. plants. A similar statement can be made in the area of suggestion systems.

What this comparison shows is that the average U.S. plant has caught up with the Japanese plants in several areas of customer, quality, and HR practices and even exceeds the Japanese in some areas. They also exceed the German, United Kingdom, and Italian plants in most of these areas of practice. Contrary to some popular opinion, things have changed for the better in the United States. As we noted earlier, manufacturing in the United States is on the rebound.

2. Strategy and Technology

But U.S. plants do not lead in all practice areas. In the middle group of practices shown in Table 10.2, U.S. plants are about equal to those in the other countries. Each country has a slight lead in one or two areas, but the lead is not overwhelming for any particular country. For example, British and German plants seem to do the best job of establishing the manufacturing-business strategy linkage. On the other hand, Japanese plants have a slight lead in areas such as formal strategic planning and overall manufacturing strategy strength. This could be attributed to the Japanese excellence at Hoshin Planning, where top-, middle-, and bottom-level goals and activities are well integrated with each other within the plant.

In the area of anticipation of new technology, no country seems to have a lead. But, the German and Italian plants lead in product design simplicity practices. With the emphasis that Germans and Italians are known to place on technology, this comes as no surprise. At the same time, Japanese plants excel at interfunctional design efforts. So, we can see that no country has a claim to strong leadership in all the areas of technology and strategy practices. Each country excels, usually by a small margin, in one practice area or another.

3. Shop Floor Practices

Here, it is fair to say that Japanese plants excel over all other countries, and the average U.S. plant falls behind on many of the comparisons. For example, Japanese plants are far ahead of all other countries on feedback given to employees. The same can be said about rewards for quality. Here, Japanese plants are using gifts, recognition, and other less tangible reward systems to a greater extent than monetary rewards.

The comparison of the Japanese lead in shop floor practices can be further extended to include coordination of decision making, shop floor contact by engineers, managers and workers, setup-time reduction, and maintenance. In almost every area that touches the workers and that matters to workers, Japanese plants not only exceed the U.S. plants but also exceed those in most other countries.

Overall Observations

This analysis seems to indicate that plants in each country excel on a certain subset of practices. No country excels at everything. In particular, U.S. plants have a lead in customer, quality, and HR practices, while Japanese plants have a clear lead on the shop floor. It is not accurate to say Japanese plants, or U.S. plants for that matter, excel at manufacturing. It depends on the aspect of manufacturing with which you are dealing.

It is one thing to discuss practices, but there is also interest in how U.S. plants compare to those in other countries in performance and what practices might lead to high performance in the United States. These questions will be addressed next.

How Do U.S. Plants Perform?

We have made comparisons of U.S. plants with plants in other countries on various performance measures. In our HPM database we measure performance in two different ways: (1) by the perception of plant management about how their plant compares to the competition and (2) by objective measures of plant performance on dimensions including cost, quality, delivery, and flexibility. In the following comparison, we have used the objective measures of plant performance. We use a combined weighted measure of cost, quality, delivery, and flexibility to rank plants from all countries in our database.

The plant rankings for U.S. and other plants are shown in Table 10.3. In this comparison we show the percentage of plants from each country that are in the top 25 percent of all plants, the next 25 percent, and so on for each quartile. We observe that the plants are fairly evenly distributed—there is no country whose plants dominate the others in every category.

Table 10.3

Comparisons of Plant Performance Using Weighted Objective Measures

	Percent of Plants Falling into Each Quartile/Percentage				
	United States	Germany	Japan	Italy	United Kingdom
Top quartile—76th to 99th percentile	48%	33%	19.5%	13%	8%
50th to 75th percentile	24	14	32	34	8
25th to 50th percentile	8	33	27	19	50
1st to 25th percentile	20	20	22	34	34

Source: HPM data.

The top quartile of performance is more highly loaded with U.S. plants, as a percentage, than any other country. Fully 48 percent of the U.S. plants are in the top quartile of all of the plants. The United States has 12 plants in the highest performing quartile, while Germany was a close second with 10 (33 percent) and Japan has 8 plants (19.5 percent). This further demonstrates that manufacturing has rebounded in the United States.

Although Germany has one-third of its plants in the top quartile, it also has one-third in the third quartile. So within the German plants, many plants are very good, and some are just below the median performance. The Japanese plants are fairly evenly distributed in plant performance, from top to bottom, in all quartiles.

We are also interested in which practices lead to higher performance, especially among the U.S. plants. Therefore, we have calculated statistical correlations among the top group of variables from Table 10.2. We display these calculations in Figure 10.2, which shows the correlations between practices and performance. Here, we indicate that top management and supervisory leadership is a driver for four different practices. Leadership is a driver in many frameworks, including the Baldrige Framework for Quality Management.

The second set of variables shown in Figure 10.2 is customer involvement, multifunction employees, small-group problem solving, and supplier quality improvement. Recall that U.S. plants exceed the level of implementation of most other plants in the world on these four variables. What is more important is the high correlations (.45 or above) between top management and supervisory leadership, as the driver, and these four variables.

Figure 10.2
Variables Leading to Performance in U.S. Plants

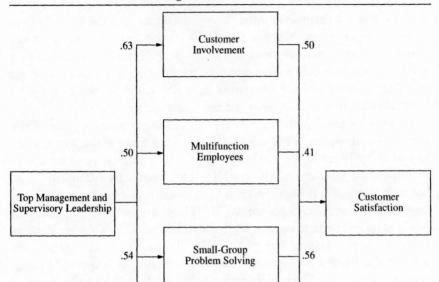

This indicates there is a strong connection between high levels of the leadership driver and high levels of the four variables. Finally, note that all four of the variables, are, in turn, highly correlated with customer satisfaction. In this case, we are seeing that the ultimate result, customer satisfaction, is driven by the four intermediate variables which, in turn, are driven by leadership. The model indicates that plants can improve in customer satisfaction, provided that top management and supervisory leadership is strong and that practices such as customer involvement, multifunction employees, small-group problem solving, and supplier quality involvement are implemented at the plant level.

CONCLUSIONS AND MANAGERIAL IMPLICATIONS

It may come as a surprise to some that manufacturing in the United States is strong once again. During the 1980s and 1990s, Americans were bombarded with constant news stories about the Japanese manufacturing miracle and failing U.S. manufacturing firms. Although this was true at the time, U.S. manufacturing companies have responded to the challenge by implementing new technologies and new management practices.

As we illustrated with Plant A, the United States now has some of the best plants anywhere in the world. We also have some of the worst, as illustrated by Plant B. Nevertheless, when we compare averages across countries, we note several things: First, U.S. productivity improvement has rebounded and is high. Second, the comparison among plants in our sample indicates that 48 percent of U.S plants are in the top quartile in performance. Furthermore, as we have indicated, this is no accident or macroeconomic effect. The plants with the best practices are able to achieve the highest performance levels. It's as simple as that. Manufacturing plants *can* achieve high performance by being diligent about selecting and perfecting their use of best practices.

The future is bright for U.S. manufacturing. To continue on the forefront will require leadership and imagination. Plants cannot stay in the lead by simply copying other plants; they must be innovative in solving management problems and in developing new technologies. In this way, leaders can retain their leadership and show the way for others.

The world is becoming increasingly a global market, with global manufacturing practices. We have seen that plants from one country cannot dominate the world market in manufacturing. There is an equalizing force that seems to restore the balance as those plants that are behind seek to find new ways to compete with and to catch their competitors. This equalizing force has come to the United States, whose plants are now, once again, competitive with any plants in the world.

REFERENCES

Schonberger, Richard. *Japanese Manufacturing Techniques: Nine Hidden Lessons in Simplicity*. New York: Free Press, 1982.

_____. *World Class Manufacturing*. New York: Free Press, 1986.

Womack, J.P., D.T. Jones, and D. Roos. *The Machine That Changed The World*. New York: Rawson, 1990.

CHAPTER 11

JAPANESE MANUFACTURING ORGANIZATIONS: ARE THEY STILL COMPETITIVE?

MICHIYA MORITA, SADAO SAKAKIBARA,
YOSHIKI MATSUI, and OSAMU SATO

During the 1980s, it seemed as though Japan was the only benchmark for manufacturing. Manufacturers around the world expressed great concern about falling behind the global benchmarks set by Japanese manufacturers. Emerging manufacturing concepts, such as just-in-time (JIT) manufacturing and total quality management (TQM), were widely accepted as the new manufacturing approach for creating competitive advantage. Japanese terms, such as *Kaizen, Poka-yoke,* and *Kanban,* even found their way into English manufacturing terminology.

Views about Japan, however, have changed drastically since then. Today, we are just as likely to find an article in the media reporting that the economic strength Japan enjoyed during the past decade has languished as the bubble economy burst in the early 1990s. Japanese business executives often argue that the source of the problem comes from structural deficiencies in the Japanese economic system, where the inefficient service sector is mixed with the relatively competitive manufacturing sector. Others argue that Japan can still be competitive because the manufacturing sector has not lost its competitiveness. While the overall Japanese economy may be struggling, many well-managed manufacturing companies such as Toyota, Honda, and Sony, still enjoy record profits. Thus, despite the sluggish overall economy, many Japanese manufacturing firms still remain competitive in the global economy. Is it just a few exceptional companies, such as these, that carry the brunt of the success of Japanese manufacturing, or is manufacturing excellence in Japan still widespread, as believed in the 1980s?

199

JAPANESE MANUFACTURING STRENGTH: HOW WAS IT ESTABLISHED?

Many believe that the strength of Japanese manufacturing organizations comes from their use of practices such as JIT and TQM, along with the operational efficiencies that have accompanied their use. We believe this is only part of the story. Manufacturing strength cannot be achieved simply by copying the practices that have made some well-managed Japanese manufacturers successful. Indeed, there are numerous examples of manufacturers in the United States and other countries that have copied practices such as JIT and TQM, with less than stellar results. What, then are the underlying forces that have made these practices so successful? We have found that there is a "communication and action" process that accompanies the use of Japanese manufacturing practices. The combination of the use of state-of-the-art manufacturing practices with the communication and action process has transformed Japanese manufacturing organizations into effective learning organizations.

How does the communication and action process work? Communication and action interact with each other, in a cyclical fashion: communication is promoted through actions, and actions are improved through communication (see Figure 11.1). Communication is enhanced when its action context is shared with people. Practices on the shop floor, for example, are a good platform for promoting communication among employees. As shop floor workers communicate with other workers about problems they have encountered on the shop floor, the actions the workers take will improve as well because they will bring more relevant information to bear on their decisions. Thus, communication and actions are mutually reinforcing. A never-ending series of dynamic and evolving communications and actions leads to the development of an adaptive capability that allows organizations to deal with their everchanging competitive environments.

The communication and action cycle has been an integral part of Japanese manufacturing management as it has evolved through three distinct stages, shown in Figure 11.2. In fact, we believe that communication and action have encouraged and enabled this evolution.

Stage I: Management Process Development

The foundation for today's Japanese management approach was established during the 1950s and 1960s. It was during this time period that TQM and

Figure 11.1
The Communication–Action Cycle

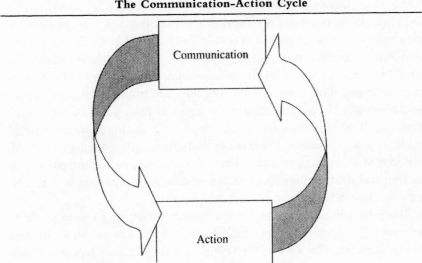

JIT started to really be developed in major Japanese manufacturing firms. In addition to using TQM and JIT, many manufacturing organizations also started to establish their own supplier networks. Initially, supplier networks were simply used as suppliers for parts; however, they gradually evolved into integrated systems for cost reduction and quality improvement.

By molding JIT and TQM into a workable approach, Japanese organizations have been able to coordinate activities through communication.

Figure 11.2
Evolution of Communication and Action in Japanese Manufacturing Plants

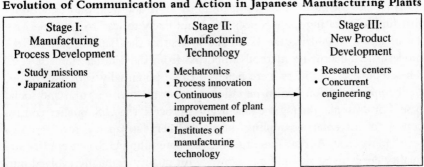

Stage I: Manufacturing Process Development	Stage II: Manufacturing Technology	Stage III: New Product Development
• Study missions • Japanization	• Mechatronics • Process innovation • Continuous improvement of plant and equipment • Institutes of manufacturing technology	• Research centers • Concurrent engineering

Along the way, Kaizen (continuous improvement) activities have been embedded into this management approach. This evolution of approaches has led Japanese manufacturers to realize the importance of integration and synchronization of various aspects of manufacturing. It was in Stage I that they started to sense that communication within and outside of the organization was a key factor for the successful implementation of manufacturing systems. The foundations of the communication and action management style were developed through this process, and Kaizen activities played an important role in developing the communication network within the organization. The communication and action management style also played a very important role later on when Japanese manufacturing organizations shifted their focus to manufacturing technology and then to new product development.

It is interesting to note that some of the manufacturing approaches often known as "Japanese" did not originate in Japan, at least not in their original conception. The sources for several of the so-called "Japanese manufacturing management" concepts were the Japanese study mission groups sent to the United States during the 1950s and the 1960s. The magnitude and number of these missions is impressive. The Japan Productivity Center reports that, starting in 1955, there were 4,464 such missions sent to the United States over a 10-year period. A typical study mission was comprised of members from industry, academia, and government. They generally focused on U.S. manufacturing concepts, such as industrial engineering, production management, and quality management, which were relatively new to many Japanese at that time. The many U.S. study missions to Japan over the past 15 years or so parallel the early Japanese study missions to U.S. manufacturers.

As Japanese managers tried to implement the concepts they had learned about in the United States, they realized that some of the concepts did not really fit into the business environment in Japan. In addition, they understood that simply imitating these concepts was not likely to lead to a competitive advantage. It was during the 1950s and 1960s that the "Japanization" of the modern manufacturing methods imported from the United States started. These concepts and ideas were believed to be potentially very useful, but their implementation had to be tailored to Japan's own cultural and social base. For example, during the Japanization process, the U.S. quality control concept of "acceptance sampling" began to emerge in Japan as the "zero defects" movement. At the same time, the then-prevailing U.S. concept that the quality department should be responsible for quality gradually evolved into the concept of "total quality management." In terms of inventory management, the U.S. concept of "buffer inventory" for avoiding production shutdowns gradually evolved to the roots of the JIT system, with a goal of removing all types of waste throughout the manufacturing process.

The Japanization process involved passing U.S. manufacturing ideas and concepts through two screens: (1) the national culture screen and (2) the organizational culture screen (see Figure 11.3). Some concepts failed to go through the national culture screen and were then abandoned. Others were modified to fit the Japanese culture. Prevailing Japanese business practices such as lifetime employment, seniority systems, and company unions helped to develop the evolving concepts because activities such as JIT, TQM, and Kaizen required people to communicate and to work together steadily toward a common goal. The result of the Japanization process was the development of a set of activities that were based on sound practices observed in the United States, but that were adapted and shaped to the unique cultural and organizational context of Japanese businesses. In contrast, as U.S. organizations tried to import Japanese practices years later, many tried to transplant them intact, without modification, often without achieving similar results.

Thus, one of the major sources of Japanese manufacturing success was that companies (such as Toyota) could discern the potential of foreign management concepts (such as quality control) but were then able to adjust and develop their own systems, adapting it to the Japanese environment. More importantly, Toyota and other organizations started to establish a process for communication and action, both within and outside of their organizations. This helped develop the capability for well-managed Japanese organizations to adapt to changes in their business environment.

Stage II: Manufacturing Technology

As the Japanese style of management began to be effective, many Japanese manufacturers turned their focus to manufacturing process innovation. It was during the 1970s that an approach known as "mechatronics" was developed in Japan. This term combined two English words: *mechanics* (mechanical engineering) and *electronics* (computer technology). The results of the mechatronics movement are seen in the emergence of robotics and other proprietary manufacturing technologies during the 1980s.

Process innovation takes place in a number of different ways in Japanese manufacturing organizations. Many Japanese manufacturers have their own

Figure 11.3
The Japanization Process

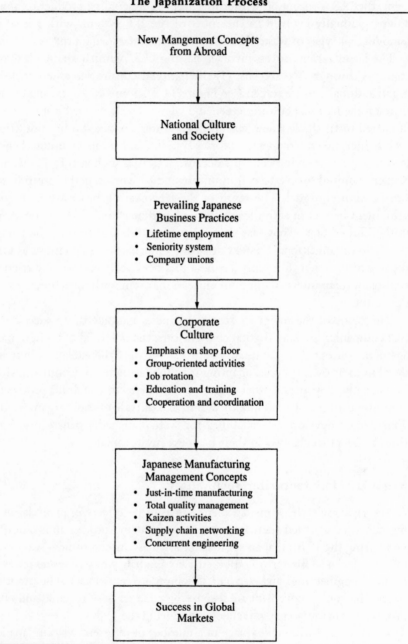

equipment engineers, who build machines that are tailored to the unique characteristics of their process environment. Other manufacturers have their own equipment firm for the group (i.e., the Toyota group has Toyota Machinery), where the equipment firm takes care of developing most of the process technologies needed by the group. Other Japanese manufacturers design their own equipment but protect its proprietary nature by using a "divide and conquer" approach to having it made. They break equipment orders down into parts, giving orders for different parts to different suppliers. As the parts arrive, they are assembled into equipment. In this way, there is no chance for process technology information to be leaked to competitors. The final approach used by some Japanese manufacturers is to buy standard equipment and then modify it to fit the unique characteristics of their manufacturing environments. These approaches all reflect the goal of Japanese manufacturers of differentiating themselves from their competitors through *proprietary process technology*.

An important feature related to process technology has been continuous improvement efforts related to plant equipment and machinery. Japanese process and equipment engineers are typically stationed on the shop floor and frequently walk around the shop floor, observing operations and communicating with operators. Through regular communication with workers, many process improvement ideas evolve into actions. Thus, continuous improvement is also a critical factor in process improvement. Again, a key success factor for the development of effective proprietary process technology in Japan is communication between engineers and workers, along with the actions resulting from the communication. When visiting Japanese factories, we found that process engineering staffs were typically housed close to the shop floor. The wall separating the floor and the process engineering staff room was transparent, made of glass. When an unusual event happened, a simple hand signal from the shop floor was enough to call the engineers. This provides an example of the close communication between people on the floor and the engineering staff and of the way in which communication can be translated into immediate action.

Large Japanese electronics companies often have their own institutes of manufacturing technology, in addition to having research-and-development (R&D) institutes. Institutes of manufacturing technology serve as a matchmaker between technological innovations and shop floor practices, that is, a device for making technological innovations competitive weapons. Institutes of manufacturing technology were established in many Japanese organizations as long as 20 years ago. At that point, many Japanese organizations had recognized the potential of manufacturing technology to be a key success factor for creating competitive advantages. Thus, real innovative capability is supported by both product and process technologies in Japanese manufacturing firms.

Stage III: New Product Development

In the mid-1980s, many Japanese firms shifted their focus to new product development. Organizations established research centers to focus on the introduction of new products or to seek technologies that led to the development of new products. In addition, the communication-based management style led to the evolution of the product development approach, which is described today as *concurrent engineering*, a process by which people from different functions of the organization work together in the early stages of product development.

The critical factor in the concurrent engineering approach is, again, communication among members of different functions of the organization, with suppliers and, in some cases, with customers. As investments were made in both R&D and concurrent engineering, Japanese manufacturers developed a relative strength in product development. Use of the concurrent engineering approach, which requires close coordination, cooperation, and communication within and outside of the organization, drastically shortened the time needed to develop new products. Information technology was also used to shorten the development period. However, the real effectiveness of using information technology is enhanced if the communication between the relevant people is effectively maintained.

Through their transition over the three stages, Japanese manufacturers developed the so-called "Japanese management style," based on communication and coordinated, cooperative activities throughout the communication network. The common theme for all three stages of Japanese manufacturing management development was the communication and action process. It is the speed of this transition that generates the competitive edge for well-managed Japanese manufacturers and simultaneously provides an adaptive capability for the everchanging business environment.

CREATION OF A LINKAGE STRUCTURE THROUGH COMMUNICATION AND ACTION

Consider the example of two Japanese plants, Plant X and Plant Y. Plant X is a medium-sized manufacturer of automotive parts. It has tried to keep at the cutting edge of manufacturing practices, implementing TQM and JIT during the middle 1980s, along with reduction of its supply base and strategic operations planning. The results have been impressive, leading to a reduction in work-in process inventory from 22 days to one day, doubled productivity per person, a sixfold reduction in space, and lot sizes that were cut from 96 to 6. Although there have been dramatic improvements in product quality, manufacturing cost as a percent of sales has decreased by 10 percent. Not surprisingly, Plant X is held up as a model within its corporation, frequently visited by the managers of other plants within the firm.

Like Plant X, Plant Y is also a medium-sized manufacturer of automotive parts. Management at Plant Y has also tried to keep up with manufacturing trends and has followed a pattern of process innovation similar to plant X, implementing JIT and TQM, reducing its supplier base, and working to develop an operations strategy at about the same time as Plant X. However, these initiatives have had little effect at Plant Y; in fact, the situation seems to have worsened, rather than improved. Although product quality has improved slightly, Plant Y's manufacturing costs as a percent of sales have increased by an alarming 12 percent. As inventory levels have been reduced, there have been numerous stockouts, leading to increased delivery times and to an increasing number of customer complaints. It seems as though Plant Y is in a rut, with little chance of digging out of it. In fact, it seems as though every time a new practice is implemented at Plant Y, the situation gets even worse!

Is the difference between Plant X and Plant Y just a matter of luck? The two plants are similar on many dimensions, and both have pursued similar innovative practices. Or is Plant X implementing those practices in a way that is different from Plant Y's approach, leading to its stellar performance? We believe that the difference is due to Plant X's ability to exploit linkages between practices through action and communication.

Patterns of Linkage Indicate "Excellence"

The importance of linkages between activities can be shown with the data we collected in Japan. Our data contains 32 Japanese factories known *a priori* as "excellent" and 14 randomly selected Japanese factories. Across all groups, there were 15 plants in the machinery industry, 16 in the electronics industry, and 15 in the transportation components industry. The "excellent" group was selected based on the market share of their main products and their general reputation. This group was further classified into two groups, based on their average score on the practices they employed as reported by representatives of each organization. The first group (group A) is comprised of the above-average half of the "excellent" group, and the second group (group B) is the remaining plants in the "excellent" group. The third group (group C) consists of the 14 randomly selected plants.

We focused on 48 manufacturing practices, which we classified into 11 management drivers, based on their similarity of functional characteristics (see the appendix to this chapter for the contents of each driver). Figure 11.4 exhibits the levels of the 11 management drivers for the three groups.

As shown in the figure, the same pattern occurs in all three groups, where the strongest drivers are the strategy, technology, performance information, organizational quality systems, operational quality systems, HR development, and shop floor. However, this pattern occurs at different levels for each of the three groups, with parallel gaps between the groups in terms of the implementation level of the practices in the three groups. The parallel gap appears when there is a positive relationship among the drivers. The higher implementation level is achieved through mutually constructive or productive interrelationships among the practices, as described in Chapter 3. A high level in one practice can be attained through the linkage of positive interactions with other strong practices, due to their mutually supportive interactions. We call this positively linked structure "levered linkage." This is the sort of situation that exists in Plant X. Its structural characteristics will be described later.

If strength in a few practices can be leveraged into strength across the board through the linkages between practices, what happens when an organization is weak in a few key practices, such as Plant Y? We found that a lower level of implementation implied the existence of a vicious cycle, where each driver constrains the others. We called this negative type of linkage a "trapped linkage." In an organization with a trapped linkage structure, each aspect or practice fails to boost others, at best, or drags them down as one aspect of manufacturing works as an impediment to others.

In between, there is a possibility of "transitive linkage," where managers strive to pull key practices upward to escape from a trapped linkage situation. A transitive linkage situation occurs when management attempts to achieve strength in key practices, making them the driving force for improving other linked practices. Another possibility is that a transitive linkage occurs when an organization is moving backward from a levered linkage situation to a trapped linkage situation. Thus, transitive linkage may serve as an early warning signal to an organization, perhaps leading to remediation.

Figure 11.4
Levels of Drivers of Three Groups

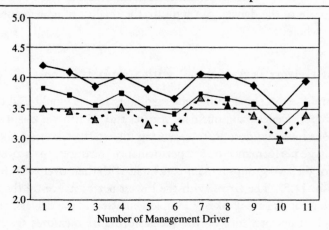

Number of Management Driver

——◆—— Above-average group of the prior "excellent group"

——■—— Below-average group of the prior "excellent group"

■ ■▲ ■ Random sample group

Drivers:

1. Strategy
2. Technology
3. Managerial
4. Performance information
5. Production planning and control
6. Production system
7. Organizational quality systems
8. Operational quality systems
9. HR development
10. Incentive system
11. Shop floor

Source: HPM data.

With respect to the three groups shown in Figure 11.4, the A group can be classified as the plants with the levered linkage. The B group is the plants with the transitive linkage, and the C group represents the trapped linkage state. Although the shapes of the three groups' patterns look the same, the underlying linkage characteristics, which determine properties of the interrelationships between the practices, show different levels of achievement of practices. Thus, when the linkage structure is known, the simple average score of the drivers can indicate the quality of management practices as a whole.

Level of Drivers Is Associated with Competitiveness

We naturally expect that well-implemented practices would lead to better plant competitiveness, and our data verifies this. When we classify the organizations into two groups, representing higher-than-average and lower-than-average performance on six performance measures, the scores for the drivers for the two groups remain consistent with their overall performance (see Figure 11.5). The firms with the lowest performance on the six performance measures are weakest in implementation of the drivers, and those with the best performance on the six performance measures are strongest in implementation of the drivers.

What about the potential for tradeoffs between measures of performance? Is it possible for an organization to be simultaneously strong on multiple measures of performance, or is excellence on one measure gained at the expense of performance on another? Figure 11.6 provides an aid to our discussion of this issue. It shows the six competitive measures for the two performance groups. We can see that it is clear that the best organizations do well on all measures of performance; thus, it is difficult to say that a tradeoff relationship exists between the competitive measures. Organizations in the higher-than-average performance group show better performance on all measures than organizations in the lower-than-average performance group. We consider the organizations that are able to excel on all measures of performance the "high performance" manufacturers.

Figure 11.6 is helpful in understanding the nature of the tradeoffs and linkages. It shows that the size of the gap between the groups is larger for some important competitive measures, such as cost and flexibility. Most of us would agree that improving delivery performance or reducing defectives is easier than improving measures like cost if an organization has ample inventory and makes thorough quality checks. However, using such an approach weakens the cost measures (a tradeoff exists). Manufacturing cost and flexibility are invisible to customers, but quality and delivery performance are directly evaluated by them. Therefore, organizations are motivated to at least improve the visible measures in order to survive competitive threats. Then the gap between such measures will be smaller, compared with those of cost and flexibility.

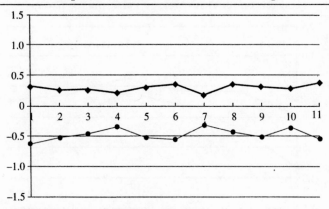

Figure 11.5
Competitive Measures for the Two Groups

—— Above-average performance group

—— Lower-than-average performance group

Drivers:

1. Strategy
2. Technology
3. Managerial
4. Performance information
5. Production planning and control
6. Production system
7. Organizational quality systems
8. Operational quality systems
9. HR development
10. Incentive system
11. Shop floor

Source: HPM data.

Escape from a tradeoff situation requires that an organization exploit the leverage effect of linkages. In other words, the organization needs to step up to the next level in the levered linkage structure. A tradeoff exists when an organization remains in the same leverage level for a linkage, but the tradeoff can be overcome if the organization can move up to a higher linkage, for example, from trapped linkage to transitive linkage or to a leveraged linkage. For example, it is possible to improve cost and quality

Figure 11.6
Performance Comparison of the Higher-than-Average Performance
and the Lower-than-Average Performance Groups

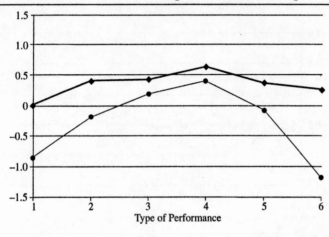

Type of Performance

——— Higher-than-average performance group

——— Lower-than-average performance group

Type of Performance:

1. Cost
2. Quality
3. Delivery Speed
4. Delivery timeliness
5. Inventory turnover
6. Flexibility

Source: HPM data.

performance at the same time if organizations can improve the leverage effect of their linkage structures. This means that management can find opportunities to realize higher levels of competitiveness, free from the peril of tradeoffs, by enhancing the leverage of their linkage structures. Cost and flexibility measures need a wider scope for changing various functions or practices extending to product design, as well as designing the alignment of various process components beyond the shop floor. More functions should be involved in the effort to improve such measures. Therefore, we can say that global competition requires us to be more cognizant of the leverage of the linkage of practices.

UNDERLYING LINKAGE STRUCTURE

Management Cycle throughout the Factory

In an uncertain situation, people sometimes resort to trial-and-error behaviors. Although they seem decidedly unscientific, an effective trial-and-error implementation can lead to top-notch results. In fact, the effective use of trial and error is a hallmark of high performance manufacturers. The trial-and-error approach is reflected in the "management cycle" concept. Japanese manufacturers have called it the Plan-Do-Check-Act (PDCA), or Plan-Do-See, cycle. This approach is used in various areas of organizations, including the shop floor, functional areas, and top management.

The most important use of the management cycle is in strategic planning and implementation. Important issues faced by organizations include how to be competitive in the current environment and how to survive in a changing environment, including changing technological developments and new product/process innovations. Many plants commit all their resources and energy to existing systems, focusing on current competitive conditions. However, this may mean that their vulnerability to change will increase, due to the increased irreversibility of decisions. Providing guidance to improve the flexibility of an organization's system to change is one of major strategic tasks for top management.

Long hours of discussion with many Japanese managers have convinced us that overall performance over time can be enhanced only through coordinated activities, both their strategic and operational aspects. However, the management cycle can be completed only if the strategic ideas are implemented properly by operational actions. Special attention must be placed on the linkage between strategic and operational aspects. Both aspects must be well linked in order to achieve satisfactory performance over time.

Linkage of both strategic and operational actions leads to satisfactory performance of a plant over time, which develops the organizational culture necessary to promote a desirable linkage structure. A positive linkage between strategic and operational actions will require certain intermediary functions or systems because they tend to be very different from each other in their focus, often contradicting each other. The strategic aspect puts its focus on future, long-range competitiveness and changes, whereas the operational aspect focuses primarily on current operations. This provides a critical barrier that most companies experience when they try to sustain their performance over time.

There are several ways in which Japanese plants link the strategic and operational aspects of the management cycle. Our research on Japanese manufacturers suggests that middle managers play a critical role in connecting the two aspects. Middle managers serve as a channel for reporting the strengths and weaknesses of existing operations to top management so that strategy can reflect a realistic picture of the plant. They engage in communication with managers of other functions to coordinate and carry out the strategic plan, with middle managers creating a realistic operational plan to achieve the goals of the strategic plan. Middle managers help their workers to prepare for changes in advance. Only middle managers can coordinate the focus of employees between current activities and activities focused on preparing for the future. Top managers are too removed from the real situation to direct employees appropriately.

Another way in which linkage of the strategic and operational aspects of the management cycle occurs is through high-level improvement teams. A team comprised of the plant manager and lower-level managers from excellent factories moves from one factory to another, performing restructuring tasks. This often does not please the people in charge of personnel allocation because the head of the team, usually the plant manager, often overrules their allocation decisions. However, the team head behaves in that way because he or she knows that the team's performance depends on closely communicated and well-organized and well-executed behaviors, especially when implementing strategic changes. Such teams play the role of a linking pin to combine the two aspects of the management cycle.

If there is an unbalanced structure, such as too much emphasis on strategic behavior or too much reliance on operational efficiency, it will be difficult for an organization to sustain its performance over time. An overly strategic focus may bring about neglect of essential weaknesses relative to current competition or may let an organization move to another business area without accumulating sufficient resources. This may be a strategic move, but it is not an excellent one if the organization is ultimately beaten by its competition. Excellent strategic management should be supported by strong operations. On the other hand, an overly operational focus can lead to problems as well, as exemplified by many historical examples including Ford's losing battle against GM after the success of the Model T Ford. Too much success relative to current competition may actually construct a barrier to new development or to changes through the complacency that results. Thus, constructive interaction between the strategic and operational actions sustains satisfactory performance over time. It is an essential condition for excellence in strategic management.

Structure of the Levered Linkage

When we think of the structure for an effective management cycle with respect to Japanese organizations, the structure shown in Figure 11.7 holds throughout all sampled companies. This structure was extracted based on the statistical relationship between practice levels. The practice drivers fell into a set of clusters, based on their interrelationship. Membership in a cluster means that a plant's component drivers' practice levels are strongly correlated with each other. If one driver is highly practiced, the others also are well practiced, and vice versa. Figure 11.7 shows the clusters of drivers, enclosed by bold lines.

The first cluster of practices is concerned with strategic issues. It consists of the strategic, the technological, and the managerial drivers. We call this the Strategic Cluster. The second cluster is the Quality System Cluster, consisting of the organizational and the operational quality system drivers. The third cluster, the Production System Cluster, is formed by the strong positive relationships between the production planning and control and the production system drivers. The fourth cluster's components are the shop floor, the human resource development, and incentive system drivers. This cluster, the Shop Floor Cluster, is concerned with quality of motivation and work of the employees on the shop floor. The fifth cluster, called the Measurement System Cluster, has one component, the performance information driver.

In addition to the strong interrelationship between drivers within a cluster (correlations between them were all in excess of .65), bold lines between clusters show strong relationships between them. The structure shown in Figure 11.7 was formed based on the relationship between individual practice levels. It indicates connections among all practices, underlying the parallel gap shown in Figure 11.4. Linking local efforts to construct linkages between the practices or the drivers enhances the level of each practice by the interaction between the practices, leading to high performance.

Table 11.1 summarizes the degree of relationship between the driver clusters of the higher-than-average performance and the lower-than-average

Figure 11.7
Structure of the Relationship between the Drivers

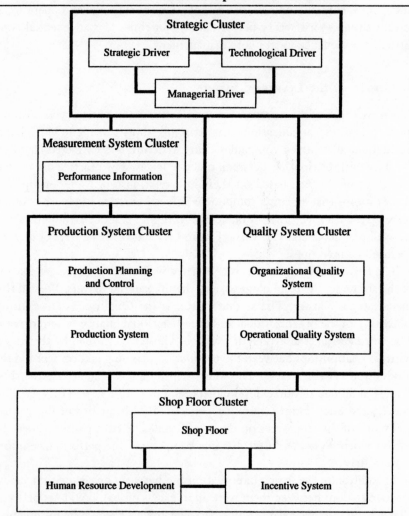

performance groups. When the degree of the relationship is high in the higher-than-average performance group, it likely means there is a leveraged linkage. On the other hand, when the degree of relationship is high in the lower-than-average group, it implies a trapped linkage because the drivers in the lowest driver clusters are dragging down all related clusters of drivers.

Table 11.1
Correlation between Clusters

	Higher-than-Average Group					Lower-than-Average Group				
	Managerial Driver	Measurement System	Production System	Quality System	Shop Floor	Managerial Driver	Measurement System	Production System	Quality System	Shop Floor
Strategic cluster	.894	.617	.737	.729	.823	.68	.866	.739	.656	.612
Managerial driver		.568	.856	.715	.823		.698	.804	.406	.768
Measurement system			.631	.402	.468			.819	.607	.739
Production system				.605	.628				.599	.786
Quality system					.898					.516

Source: HPM data.

Note: The figures are all statistical correlation coefficients; 1.0 means the corresponding clusters coincide perfectly as to their changes, while 0 implies that they change independently. All correlation coefficients are significantly different from 0, indicating that a relationship exists.

The linkage structure that emerges in the relationships among the various drivers comes from excellent management practices, particularly the effective combination of operating and strategic aspects. Unlike high-performing plants, the focus of an average factory is the creation of a system of operations under a particular operating criterion, such as cost or quality, concentrating only on the efficiency of narrowly focused operations. Knowledge and understanding of the prevailing strategic focus is poor among top managers of an average company. They tend to resist any external requirements based on their own local criteria. Not surprisingly, their efforts often do not lead to improvements in overall performance.

The strategic aspect is difficult to improve in isolation. It has to be supported by operational activities, including the combined set of Quality System, Production System, and Working Floor clusters. Lack of synchronization between strategic and operational aspects leads nowhere. For example, operational actions may fail to adjust to changes in the competitive environment, whereas strategic actions cannot move forward to a better strategy leveraged on manufacturing strengths because of poor implementation of operational actions. Thus, an organization like Plant Y is constrained, which leads to a trapped linkage structure.

Strategic and Operational Aspects Are Equally Responsible

Middle manager competence is the key factor for generating a positive interrelationship between strategic and operational aspects. But how is this achieved? If there is tradeoff friction between the two aspects, their missions can be terribly complex and frustrating.

The way in which the interrelationship is dealt with is through a combination of continuous improvement with strategic leaps. Figure 11.8 shows this process. For example, after a new product or process is introduced in the strategic aspect, the operational aspect focuses on continuous improvement and achieves performance improvements, such as cost reduction. Operational efforts often face saturation as they continue, and the rate of improvement may start to diminish. When an organization's performance starts to decline, the feeling of achievement begins to wane, often leading to a situation in which apathetic attitudes start to prevail. To avoid this diminishing effect, a strategic organization would prepare to introduce its next generation of products or processes (the next strategic leap) to provide new opportunities for operational efforts and to keep the momentum going.

Figure 11.8
The Reciprocal Process of the Strategic
Aspect and the Operational Aspect

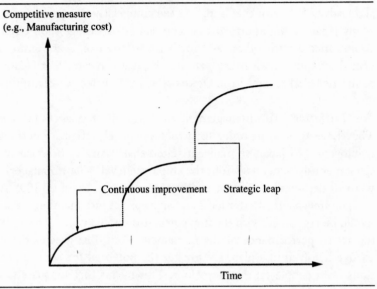

We asked Japanese managers and engineers in charge of introducing new products or processes at several major automobile companies what they expected the contribution of operational efforts to the achievement of cost goals to be. Their responses were consistently "about 50 percent." Although this figure may not be precise, their answers tell us that they all felt that the strategic and the operational aspects are equally important in achieving and sustaining a competitive position.

Communication Structure for Creating a Levered Linkage

We have focused on *actions* in the previous section, with good reason. The overall excellence of practices is a sign of a well-managed plant, and true excellence results from the levered linkage of practices. However, the quality with which practices are implemented is ultimately determined by people because the interactions that create leverage are carried out by people. This implies that only organizations with an effective communication structure that drives people toward such interactions can generate linkages between practices. Thus, we return to the communication and action cycle.

In this section, we will identify key communication factors for creating constructive communication in a levered linkage structure.

Effective communication allows people to share necessary information and leads to behavior that improves the competitiveness of an organization. Many Japanese high performance companies encourage their employees to communicate with others and to then act based on that communication. Our discussions with managers and engineers at excellent Japanese companies revealed several basic factors relevant to effective communication:

- *Initiative.* Top management and first-line managers take initiative. They have to lead in order to invoke genuinely effective actions among employees. No Japanese manager stated that desirable practices were ever spontaneously generated from the employee level. Their managerial efforts were all designed and intentionally implemented, based on PDCA cycle.
- *Improvement.* Better ideas and appropriate judgment on the shop floor should be combined with the implementation of plans. Such ideas determine the actual performance of the implementation. The greater the extent of change involved in a plan, the greater the potential for complementary actions from people on the shop floor. This means that adequate freedom of action should be given to employees. Excellent initiatives are necessary, but the ultimate goal is to encourage the creative capabilities of all employees.
- *Feedback.* Confirmation of results generates the energy for the next round of communication. In order for initiatives to successfully motivate communication, the results of the implementation should be fed back, and the meaning of their communication behaviors should be confirmed. Success is the most powerful driving force for constructive communication.
- *Creation of a culture of communication.* The most important factor is to inspire communication by success. Communication with people in other functions or departments is not always easy. Meaningfulness is required to maintain such communication. If a communication culture exists, people will see communication efforts as routine, and the time required for effective communication will be reduced. The momentum of effective communication is stimulated by success.

A series of successful results will generate a desirable communication culture. Each employee should be involved in the communication process to learn how to improve himself or herself through communication and to develop new insights for more effective actions. If they can experience the effectiveness of the process, it will be embedded into each employee's value system.

Success starts with smart ideas and well-coordinated actions. Success over time is possible through excellent management, that is, a timely and effective mix of the strategic and operational actions. The essence of constructing a levered linkage structure is to associate communication activities with final successful outcomes. Chatting without purpose is not sufficient; however, purposeful and effective communication should be encouraged. Such communication should be developed as part of the effective business process leading to satisfactory performance.

CONCLUSIONS

To date, the emphasis of reports on Japanese manufacturing management has been on manufacturing concepts such as JIT and TQM. However, the most important characteristic of well-managed Japanese manufacturing organizations is their ability to created a leveraged linkage structure throughout the communication network. This is the real source of strength for Japanese manufacturers, not the techniques themselves.

A leveraged linkage structure implies high flexibility and rapid adjustment to changes. The communication structure associated with a leveraged linkage structure makes it easier for all members of an organization to effectively pursue a common goal. Once a decision is made, people can move quickly and cross-functionally to coordinate their activities with each other in accordance with strategic aims.

So-called Japanese manufacturing management concepts such as JIT and TQM are the result of many activities and actions through the linkage structure. The concepts themselves are not difficult to copy; however, it is not easy to develop an organization with the necessary linkage structure that fits with its competitive environment, as shown by Plant Y's experience.

The majority of managers we interviewed during our research know exactly what they should do next to remain competitive. They remain confident. However, there are signs that Japan could follow the same path that many advanced nations have taken. Rising prosperity has increased the desire of young capable workers to enter the more glamorous service sector, and working in a factory is increasingly less attractive for the younger generation. During the late 1980s and 1990s, one of the choices for manufacturing organizations was to move part of their operations overseas. The linkage structure of Japanese organizations was developed in a homogeneous society, where people are more group oriented. There is no guarantee that a similar linkage structure can be created on foreign soil. Indeed, although Japanese transplants have been producing reasonable profit levels in Asia, many are struggling in North America and Europe.

At this point, we don't know whether they are struggling because of the difficulties of applying linkage-based management on foreign soil or simply because of their short history and lack of experience in overseas operations. Therefore, whether Japanese manufacturers can develop similar management processes in different environments remains to be seen. The future success and survival of many Japanese manufacturers is dependent on their ability to adapt their management style to different and changing environments. We believe that well-managed Japanese manufacturing organizations will be capable of handling the challenge.

APPENDIX

DRIVERS

1. *Strategic:* How well the factory performs strategically.
   * Formal strategic planning
   * Communication of manufacturing strategy
   * Manufacturing-business strategy linkage
   * Manufacturing strategy strength
2. *Technological:* How effectively the factory takes advantage of technological opportunities.
   * Product design simplicity
   * Interfunctional design efforts
   * Anticipation of new technologies
   * Effective process implementation
3. *Managerial:* How well middle management stimulates cooperative and interactive actions among employees and managers.
   * Functional integration
   * Coordination of decision making
   * Supervisory interaction facilitation
   * Shop floor contact
4. *Performance Information:* How adequately the factory uses data about its operations.
   * Dynamic performance measures
   * Performance feedback
   * Internal quality information
   * External information: supplier quality control
5. *Production Planning and Control:* How well the factory establishes a system for operations.
   * Stability/predictability of short-term production
   * Daily schedule adherence
   * Repetitive nature of master schedule
   * Benefits of information systems
6. *Production System:* How efficiently the factory achieves operations.
   * MRP adaptation to JIT
   * Equipment layout

- Setup-time reduction
- Just-in-time delivery by suppliers
- Kanban
- Just-in-time link with customers

7. *Organizational Quality System:* Supportiveness of the quality environment.
   - Supplier quality involvement
   - Quality of new products
   - Customer involvement
   - Top management leadership for quality

8. *Operational Quality System:* How well the factory achieves quality control on the shop floor.
   - Process control
   - Customer satisfaction
   - Continuous improvement
   - Feedback

9. *Human Resource Development:* How aggressively the factory develops human resources for operations.
   - Multifunctional employees
   - Task-related training for employees
   - Recruiting and selection
   - Manufacturing/human resource fit
   - Documentation of shop floor procedures

10. *Incentive System:* How adequately the factory rewards its employees.
    - Rewards/manufacturing coordination
    - Compensation for breadth of skills
    - Incentives for group performance
    - Rewards for quality

11. *Shop Floor Motivation:* How well people on the floor are motivated.
    - Commitment
    - Small-group problem solving
    - Employee suggestions
    - Maintenance
    - Cleanliness and organization

CHAPTER 12

FROM COMPLACENCY TO COMPETENCE: LESSONS FROM THE UNITED KINGDOM

CHRIS VOSS and KATE BLACKMON

$\mathbf{H}$ow can manufacturing organizations survive, let alone win, in today's dynamic and competitive markets? Central to developing the competitiveness of any operation is the core of continually developing practice that is high performance manufacturing (HPM). What constitutes high performance practice is surprisingly well known—the same books and *Harvard Business Review* articles are on the shelves of the operations vice presidents and managers of most organizations. However, their ability to translate this knowledge into embedded practices and high performance is remarkably variable, as is the speed at which it is accomplished.

The United Kingdom provides a particular example of a country in which this is an issue and from which important lessons can be learned. Because of both the English language and the presence of a very large number of Japanese and U.S.-owned manufacturing sites, U.K. managers are well exposed to the concepts of HPM. Despite the need to be at the cutting edge, however, few U.K. companies have both leading-edge practice and leading-edge performance—a recent study found that only 2.3 percent of the companies in the United Kingdom had achieved the superior levels of practice and performance that might be classified as world class.[1] A vital leadership task is the development of an improvement strategy to bring the organization up to tomorrow's standards of practice and performance. We can learn much from the United Kingdom about the issues of moving to high performance levels. This chapter will draw on U.K. experience, but it will also add relevant examples from other countries.

Too many managers believe that their organizations are high performing in terms of practice and performance, despite weak performance. As a result, they are complacent. Even organizations that have made a real effort to adopt a wide range of best practice find that they still lag the performance of the best in their industry. There are two key questions that executives must ask themselves. (1) Do we really know how the levels of practice and performance in all our plants rate against world class standards? (2) If high performance practice leads to high performance *and* if my managers know what constitutes high performance practice, why aren't we getting there?

The natural response to the need to continually develop the processes and the capabilities that are needed to sustain long-term success is embarking on programs of practice development. The past 20 years has seen a wide range of sets of practices and processes that have been shown to lead to improved performance. These range from lean production and the lean enterprise, to total quality management (TQM) and high performance work teams, and include techniques as varied as quality function deployment and total productive maintenance. Many of the exemplar companies, from Toyota with lean production to Motorola with Six Sigma, have had their business success associated with distinctive management practices. Another example is General Electric (GE) with its boundaryless company—seeking and adopting the best from anywhere. Yet attempts to emulate these companies have frequently led to disappointment. As the executive who arrived at one U.K. organization and found it buried in incomplete programs said, "It was death by a thousand initiatives." Does this sound all too familiar?

Lack of knowledge is a major barrier to a company's recognition of the need to improve. This chapter presents a framework for positioning a company's manufacturing practices and performance relative to competitors. It argues that a company's improvement strategy should be largely determined by its starting point, the past and the current sets of practices, and its actual performance. Using examples from the United Kingdom and other countries, we will show how successful improvement trajectories are developed from different starting points. On the one hand, complacency was turned into a fast-track journey to high performance, while on the other hand, initiative overload was turned into an effective and coordinated improvement program.

THE COMPLACENT ORGANIZATION

Complacency has been identified as a leading problem in the United Kingdom and in other countries, in both the manufacturing and the service industries. Having an outstanding product or a killer application that generates outstanding growth and profit can mask the fact that the plant may not have the processes and practices in place to cope with that sustained growth or to generate a stream of products and applications needed in the longer term. Too often, organizations see financial performance alone as an indicator of competence. Such organizations are prone to complacency. Senior management's perception of the practices, capabilities, and performance of their organization is often strikingly at odds with what a visitor can see on a tour of the shop floor. We found this particularly true of poorer-performing organizations. It is not uncommon to be told that an organization is approaching best in class, but to find good practice not in place. Where good practice was in place it was not fully deployed across the organization, and the resulting performance was well off the competitive pace. We find this to be particularly true of poorer-performing organizations. Such failure to recognize reality can be called, at best, complacency, and, at worst, severe

Figure 12.1
Relationship of Complacency to Performance

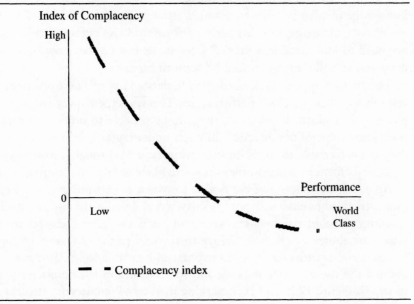

overoptimism. However, as we show later, complacency is not necessarily the exclusive domain of the poorer-performing company.

A complacent company is one whose perception of its competitiveness is higher than actual reality. A complacent organization is poorly set up for competing in the long term because it has inadequate levels of practice or is unable to translate these into performance; yet it may still think of itself as highly competitive.

The reverse of complacency is realism. Realistic companies are those that have a clear idea of both their strengths and weaknesses and their positioning relative to competitors. Complacency can be measured by comparing an organization's view of its competitiveness with the underlying drivers—the actual levels of practice and performance achieved by that organization. We have constructed a complacency index. The larger the gap between the perception of competitiveness and its actual practice and performance, the higher the level of complacency. "No gap" indicates realism, while a "negative gap" indicates pessimism. The relationship that we have found between the level of complacency and performance is illustrated in Figure 12.1 on page 227.

DRIVERS OF COMPLACENCY

The most common driver of complacency among companies is lack of knowledge of what is being achieved by others. This is typified by a statement from a manager at one poorly performing U.K. organization: "I do not need to look outside, I know." Clearly, he had no idea how his company was actually doing, nor did he seem to care.

The best companies look outside to understand what "best practice" is and what is "best-in-class" performance. This leads them away from complacency to realism. In addition, they are better able to improve through their knowledge of best practice; through understanding their weaknesses, they are better able to focus on improvement; and through knowledge of high performance standards, they are more likely to seek challenging goals.

An excellent indicator of the degree to which an organization seeks relevant knowledge externally and is outward looking is the use of benchmarking. One of the best-known examples is the Xerox Corporation, which attributes much of its ability to respond to the challenge of their Japanese competitors to its aggressive use of benchmarking. We have examined the use of benchmarking and its relationship to complacency, as shown in Figure 12.2. As benchmarking increases, complacency decreases.

Figure 12.2
Complacency, Learning, and Performance

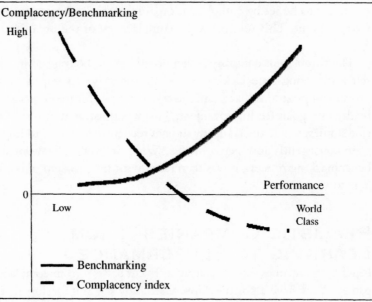

COMPLACENT COMPANIES—THE
NEGATIVE LOOP OF PERFORMANCE

Complacent companies are locked into a cycle of poor practice and poor performance. This is a negative loop consisting of a number of self-reinforcing elements.

• *Low incentive to improve.* Because they do not know or recognize their company's true position, managers of complacent companies may have less incentive to initiate improvement activities.

• *Random improvement activities.* The lack of understanding of their company's strengths and weaknesses makes it difficult for managers to make a choice among different improvement programs; as a result, their improvement strategy often consists of a random selection of improvement programs.

• *Lack of knowledge of how to improve.* Because they tend to be inward looking, managers in complacent organizations have less knowledge of the detailed content of best and appropriate practice, limiting their ability to improve.

- *Undemanding goals.* The levels of performance set by high-performing organizations are unknown; and as a result, stretch goals are less likely to be set by complacent organizations, leading to lower rates of improvement. This reinforces the complacency of the organization.

The complacent company is therefore likely to be trapped in a negative, self-reinforcing loop lacking drive and incentive to improve, with improvement programs that are not necessarily right for the organization, and inadequate goals for improvement. This was illustrated in one of the U.K. case studies, where the company did not recognize any need to improve, despite consistently low performance. When it eventually went through a benchmarking process, it was then faced with the problem of not knowing *how* to improve.

REALISTIC COMPANIES—FROM LEARNING TO PERFORMANCE

Realistic companies show a sharply different pattern from complacent companies. They have a positive loop of strongly self-reinforcing elements:

- *Knowledge that they must improve.* High-performing organizations invariably have a drive to improve. This comes from the knowledge that there is always someone doing something better; when both peers and followers are actively innovating, doing nothing will lead to falling behind.
- *Focused improvement programs.* The knowledge of strengths and weaknesses leads to intelligent focusing of improvement programs. Focusing managerial resources on a limited set of activities will lead to better improvement, and focusing on the most relevant areas will lead to superior performance.
- *Expertise in improvement.* Outward-looking organizations are more likely to understand in detail the content of new practices, in particular, their appropriateness and implementation issues. They will be aware of industry best practice as well as of practices that can be transferred from one part of the organization to another and from outside the industry.
- *Stretch goals.* Knowing best-in-class performance in each area of an organization's key processes will lead to the adoption of stretch goals for improvement—goals that simultaneously stretch the organization and are achievable.

There is a sharp contrast between those organizations that are inward looking and complacent and those that are outward looking and realistic. Not only do the best companies lead the poorer ones, but they are also able to improve faster and better. Consequently, the gap between the best and the rest will widen, not close! (See Table 12.1.)

FOR IMPROVEMENT—THE STARTING POINT IS CRUCIAL

Given the objective of reaching world class standards of practice and performance, managers are faced with a daunting array of potential areas for improvement and with tools and approaches for improving practice and performance. It is not unusual for a company to have dozens, if not hundreds, of improvement activities taking place simultaneously. Too often, individual improvement programs or tools are presented in terms of a single end point (high performance) and a uniform journey (adopting this program or tool) to achieve this end. Yet any traveler knows that the route that you take tomorrow in a long journey depends on where you have come from—today's starting point and the chosen destination. Indeed there may be a choice of different routes, depending on the number and the sequence of necessary steps on the way.

Where an organization is on its journey of improvement can be represented by its location on the dimensions of practice and performance: the degree to which it has in place the necessary practices and process, and how

Table 12.1
The Realistic versus the Complacent Company

The Realistic Company	The Complacent Company
Understands own strengths and weaknesses.	Has no view of key strengths and weaknesses.
Looks outside for new knowledge and benchmarks of performance.	Is inward looking.
Knows content of industry best practice.	Is unaware of best practice.
Has improvement strategy tailored to strengths and weaknesses.	Has random improvement programs—no coherent improvement strategy.
Sets stretch goals based on world's best.	Has goals based on historical performance.

well these have been translated into operational performance. Typically, the pattern of the practice and performance of different companies follows the pattern shown in the scatter diagram in Figure 12.3.

UNDERSTANDING THE INDIVIDUAL IMPROVEMENT AGENDA—THE "FOOTBALL DIAGRAM"

Looking at averages neglects both the individual company and the enormous spread of practice and performance in each area and country. For the individual organization, it is not this pattern that matters, but their own position in relationship to other organizations: their starting point for improvement.

At the level of the individual company, there is no single prescription for reaching high performance levels: the agenda will depend on the company's position. Companies begin at very different starting points; some have high levels of practice but have yet to realize performance. Others have good performance despite having limited practice deployment. Many are in the middle, implementing change programs but still having a considerable journey

Figure 12.3
Distribution of Practice and Performance

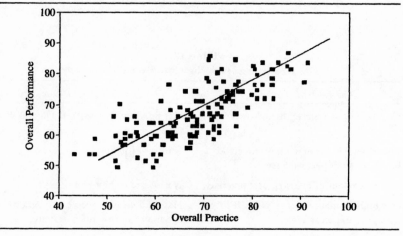

ahead to achieve the desired practice deployment and levels of performance. A company that is already at or close to HPM levels will have a very different agenda than one that is poor in all areas.

We can use the pattern of companies on the scatter diagram (Figure 12.3), which resembles the shape of an American football (or English Rugby ball), to categorize five different starting points (see Figure 12.4). To develop an improvement strategy, a company must first understand its improvement agenda. Each of these five starting points on the journey to achieve high-performing competitiveness reflects different internal or external issues facing the organization, and each has a particular agenda for improvement. Within each group, sites may share similar problems and opportunities in achieving world class practice and performance.

The Laggards

Companies in this group are those that have both poor practice and poor performance. They are the most vulnerable of all the companies. Their key dilemma is, "Given that we have so much wrong, where do we start?" Invariably, there is advice coming from many quarters. Consider, for example

Figure 12.4
The Five Starting Points for Improvement

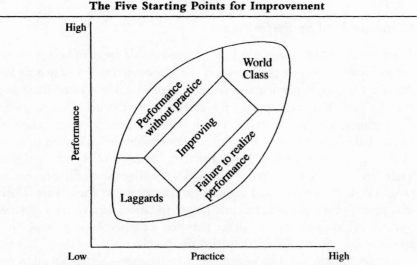

the application of computers, total quality management (TQM), just-in-time (JIT), further investment and adoption of ISO 9000. There is a multitude of performance improvement programs open to them. Choice of direction is not an easy task when there is little resource or managerial time available. Experience has shown that there are a number of guidelines for a company in this situation. First, it should make sure that the foundations for good manufacturing management are in place. In particular, all aspects of good organization should be established, and following this, quality management should be implemented. Without this beginning, investments in other programs may fail. In addition, high quality and good organization positively leverage productivity and other areas of performance. Companies in this category frequently seek computer solutions to problems that are best approached by other means. Computerizing a poor plant will not improve it. The choice of where to invest limited management resources and attention to the needs of the market are crucial. A thorough understanding of customer needs in the various markets in which the company operates can guide the choice between various programs for practice improvement. A typical U.K. case is Metal Products.[2] This company had a cluttered shop floor, an uninvolved workforce, low levels of learning, and a lack of focus. In order to find its way out of this mess, it used diagnostic benchmarking to identify the key areas for improvement so that it could launch a successful, focused program of improvement.

Failure to Realize Performance

There are many companies that have adopted a wide range of best practices but still have below-par performance. We propose a number of reasons for this. First is poor implementation of new practices. The United Kingdom has relatively more companies in this category, possibly an indicator of the thoroughness of other countries, which leads to better implementation of practices they choose to adopt. Second is uncoordinated adoption of multiple programs (often called "alphabet soup" because of the many three-letter acronyms). This is closely linked to weak alignment of improvement programs with corporate and manufacturing strategy. A final cause is lack of coherence between practices and failure to tailor them to the company context. HPM practices will differ between a high-volume process food producer and a low-volume batch aircraft assembler. One of the U.K. case companies had 197 separate improvement initiatives—clearly too many for any to be successful.

Performance without Practice

These companies have far better performance than their practices would lead one to expect. There are a number of reasons for this. First, they may have very demanding customers, who require high levels of delivery, quality, and/or low cost. However, their customers may pay little attention to how this is delivered. Thus high delivered quality may be achieved by high levels of inspection or rework, or high delivery responsiveness may be achieved by inventory. Second, outstanding product design may lead to high market performance without all around good practice in operations. The companies know how to win orders, but they do so with less than best practice. Illustrations of this include the food industry, where high levels of performance are often achieved without matching high levels of practice. This may be due to the very strong position of supermarket chains in the United States and many European countries. The challenge for companies in situations such as this is to realize their vulnerability when they are enjoying success—to overcome complacency.

Improving

The bulk of the companies in all countries fall into this category. They are adopting many aspects of best practice and may already have some in place. They will probably have met and addressed the problems that companies in the preceding groups have faced—deciding where to start and where to allocate scarce resources, developing market-led manufacturing strategies, and realizing the need to improve. They all still have a long way to go. In this category, the first challenge is to overcome inertia. Change, much less continuous change, is difficult; and in most companies, there are strong forces of inertia that may slow down improvement efforts. For example, in the quality management area, companies can adopt ISO 9000 and then rest on their laurels rather than use it as a platform for continued quality improvement. As many companies know, major programs of change, such as total quality management, can lose their momentum or can be side-tracked by short-term problems. Maintaining the momentum toward high performance levels is a key task for management.

World Class

Despite the indications of the previous examples our case research found high performing operations in a number of organizations. One such company,

"Electronic Products Limited," was characterized by a passionate desire to learn. It supported many learning initiatives, from structured benchmarking to a global intranet, where shop floor improvements could be instantly shared with any other plant around the world. For companies such as this, at or close to current world class standards of practice and performance, the challenge is to find new paths to the future, to become innovative. Unlike companies further away from HPM, these companies have no models of the future for which they can aim. These companies must find their own paths to future improvements in manufacturing practice and performance. They know that their competitors will be doing this. For example, in countries with a strong currency, many companies look for step change (30 percent or more) reduction in manufacturing cost. To do this, they must seek new

Table 12.2
Comparison of the Five Starting Points

The Position	Why Are We Here?	What Is the Improvement Agenda?
Laggard	Customers are not demanding enough.	*Starting Point* Choosing from the wide range of things that have to be done.
Performance without practice	Complacency.	*Overcoming Complacency* Generate corporate understanding and acceptance of our true position.
Failure to realize performance	Death by a thousand initiatives. Lack of internal and external coherence of improvement programs. Poor implementation skills.	*Implementation and Direction* Linking improvement strategies to business strategies. Developing a limited, coherent set of improvement initiatives. Focusing on implementation issues.
Improving	We know what to do, but we are not there yet.	*Maintaining the Momentum* Selecting the vital few each year.
World class	Dedication and commitment.	*Staying in Front* Innovation, developing tomorrow's world class practice and performance

methods and question today's established wisdom. New approaches will be developed and tried, some will fail, but others may succeed and new standards for world class practice and performance will be set.

The positions, background, and agendas for each starting point are summarized in Table 12.2. This is also illustrated on the football diagram in Figure 12.5.

FROM IMPROVEMENT AGENDA TO IMPROVEMENT STRATEGY— THREE CASES

How does a manager bring the company up to tomorrow's standards of practice and performance? This section shows how companies can get from the improvement agenda described previously to an improvement strategy. An *improvement strategy* is a choice of improvement activities that lead to an improvement trajectory over time (Figure 12.6).

Figure 12.5
Improvement Agendas for Different Starting Points

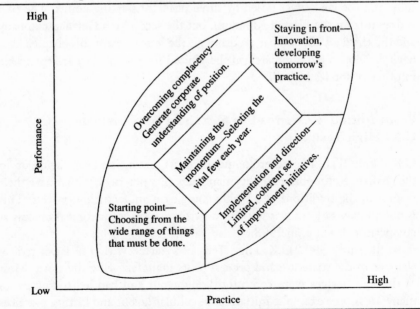

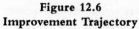

Figure 12.6
Improvement Trajectory

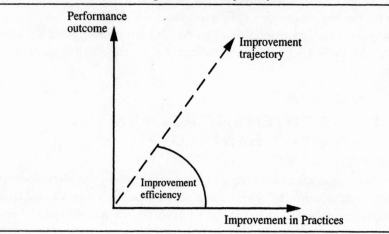

The effectiveness of an improvement strategy determines the level and the rate at which the practices lead to improved performance. As stated previously, the improvement trajectory will be governed by the company's starting point. Three case studies illustrate how companies have successfully met the challenges posed by three different starting points. The first is drawn from the United Kingdom, but the second is a German company and the third a U.S. company. Although the lessons from this chapter have been developed from research in the United Kingdom, they are applicable in other countries as well.

From Initiative Overload to Sustained Improvement—U.K. High Tech Ltd.

U.K. High Tech, like other companies in its sector, has been facing up to the challenge of modernizing its manufacturing practices, which have their origins in the traditions of a highly complex manufacturing process. This is not an easy task, as was witnessed by the major production problems at competitor, Boeing, in 1998.

In the mid-1990s, U.K. High Tech had started with a series of radical changes in the structure and process of its manufacturing divisions. Most of these initiatives were carefully thought out and had led to success in many areas. For example, initiatives in cellular layout and kitting practices had led to a reduction in the number of orders being processed on the shop floor for a given level of output—a drop to about one-tenth of the previous level. However, in too many cases, these initiatives have led to neither effective deployment of new practices nor the desired performance outcomes being reached fast enough. For example, continuous improvement, although being widely implemented, was not felt to be "embedded" in the organization. As one manager stated, "Embedding new practices is like 'rolling a boulder up the hill'—when the organization relaxes, the boulder rolls back down to its starting point!"

On taking over one of the key divisions of U.K. High Tech, the new site director found that there were 197 improvement initiatives in place! Not surprisingly, people in the organization talked of "initiative overload" and "initiative fatigue." These initiatives were fragmented and too often unlinked and unrelated to strategy. As a manager stated, "If someone has seen something good, a pilot will get started. But we rushed in without a strategy of knowing how to do it properly." The result was that initiatives frequently failed for a wide range of reasons—from lack of resource to lack of commitment or skills.

The key task was to bring these multiple initiatives together without dampening people's commitment and enthusiasm. The first step was to move from seeing improvement as a series of initiatives to seeing them as an agreed journey. A clear structure was developed around seven core processes within the organization. All improvement activities had to be related to these seven processes, which immediately created a strong focus. Three standard steps were defined: (1) focus, (2) strategy, and (3) action. A key role of strategy was that all potential improvement initiatives were reviewed against strategy before being accepted. From strategy flowed appropriate resourcing and action plans. This framework helped people position their actions on this journey. A policy deployment approach was used, whereby focus, strategy, and action are now cascaded down the organization so that both responsibilities and the relationship to strategy are clear. The next step was to cement the process in place, managers' and supervisors' responsibilities were clearly defined. In each team or area, one person was responsible for one process so that individuals were not overloaded (not all areas were involved in all seven processes). This provided visibility of responsibility, from the shop floor to director, and simple reporting relationships, but with strong ownership.

Any new process, or way of working, does not just need to be used in islands of excellence; it needs to be consistently deployed across the whole organization. To address the issue of deployment, a set of common tools was developed. To support this, a measurement system was developed to measure the progress of each process in each area along the agreed journey. This measurement system focused on the extent of deployment of practices or enablers. This was supported by internal and external benchmarking.

The net result has been a move from multiple fragmented initiatives to seven coordinated journeys to excellence. Ambitious targets for operational improvement are now being met; and, most important, all employees in the organization are now pulling in the same direction.

From Complacency to World Class—the Case of Mercedes Benz

Today, Mercedes Benz automobiles have moved from just a leading luxury carmaker to one of the leading innovators, with exciting new vehicles being launched. In the late 1980s this situation would have been difficult to predict. In the early 1980s, Mercedes Benz considered itself the best car company in the world. It dominated the luxury car market. Its models had a reputation for elegance and reliability. However, by the mid-1980s, Toyota had successfully launched the Lexus, thereby challenging all established luxury car manufacturers.

It soon became clear that Mercedes's position was not based on broad-based, leading-edge manufacturing practices. For example, its product development processes were slow, and quality was too often inspected in rather than built in (at that time Mercedes was reported as using as many man–hours rectifying faulty cars as Toyota used in assembly). Although the result was high delivered quality and customer satisfaction, it was achieved despite poor practice. Toyota, by applying its manufacturing and design process expertise to the luxury car segment, had been able to match or to exceed Mercedes's design and quality and to do so at much lower cost. However, unlike many other companies that found themselves in this situation, Mercedes challenged its own historic complacency and set about becoming a world class manufacturer, not just in the traditional strength of German engineering, but in all aspects of manufacture, from the product development process to manufacturing.

The end result was both a resurgence in its traditional luxury car segment, and a surge of innovation that threatens to take other U.S. markets by storm. These have included the sports-utility vehicle, the M class, the SLK two-seat roadster, and the CLK luxury coupe. The Mercedes results show clearly in the marketplace with a growth in U.S. market share from 5 percent in 1991 to 9 percent in 1997. Also, U.S. sales growth of Mercedes has far exceeded other brands as shown in Table 12.3. The latest chapter in this saga was the merger of Daimler Benz and Chrysler. If the lessons from this are applied across the new enlarged group, it will lead to powerful results.

Leadership and the Relentless Drive to World Class— General Electric

Much has been written about General Electric (GE) and Jack Welch; one of the reasons is that GE is a benchmark for managing the journey to HPM. We have illustrated two dangers in the journey to HPM: (1) complacency coupled with poor practice, and (2) a focus on practice often involving initiative overload and failure to realize performance. GE's strategies illustrate how both dangers are minimized. Under Jack Welch, GE has clearly avoided complacency. The drive to be number one or number two in each industry has forced a focus on performance and on benchmarking against the world's best, seeking to surpass them. The concept of "boundaryless behavior" has been at the heart of this effort.

The sweetest fruit of boundaryless behavior has been the demise of "Not-Invented-Here" and its utter disappearance from our company. We quickly began to learn from each other: productivity solutions from Lighting; "quick response" asset management from Appliances; transaction effectiveness from GE Capital; the application of "bullet-train" cost-reduction techniques from Aircraft Engines; and global account

Table 12.3
Sales Growth of Mercedes in
Luxury Car Market

U.S. Sales Growth of Luxury Car Market 1991–1997 (Selected Brands)	
Mercedes	83%
Lexus	28
Cadillac	−14
Lincoln	−22
Acura	−26
Overall market	+6

management from Plastics—just to name a few. At the same time, we embarked on an endless search for ideas from the great companies of the world. Wal-Mart taught us the direct customer feedback technique we call Quick Market Intelligence. We learned New Product Introduction methods from Toshiba, Chrysler, and Hewlett-Packard, and advanced manufacturing techniques from American Standard, Toyota, and Yokogawa. AlliedSignal, Ford, and Xerox shared their insights into launching a quality initiative. Motorola, which created a dramatically successful, quality-focused culture over a period of a decade, has been more than generous in sharing its experiences with us. . . . We embarked on an endless search for ideas from the great companies of the world.[3]

GE has taken the study of other high-performing organizations and embedded it in its organization. However, Welch has been very successful in avoiding the pitfalls of the "practice without performance" group. For years, he was skeptical of the quality programs that were the rage of the 1980s; he felt that they were too heavy on slogans and too light on performance. However, when the programs were proven to yield major performance outcomes in the GE context, he rolled them out aggressively. An example is Six Sigma. Lawrence Bossidy of AlliedSignal, Inc., convinced GE that the program that had been highly successful for him could be the same at GE. As a result GE launched a major effort in 1995, with major projects and intensive training programs.

"We didn't invent Six Sigma—we learned it. Motorola pioneered it and AlliedSignal successfully embraced it. The experiences of these two companies, which they shared with us, made the launch of our initiative much simpler and faster." This program expanded to 3,000 projects in 1997. In 1997, the program delivered over double the planned productivity savings.[4] In 1999 the Six Sigma program produced more than $2 billion in benefits.[5] Performance orientation has been hand-in-hand with best practice orientation. Welch sets stretch targets and drives managers to achieve them: "in stretching for these 'impossible' targets, we learned to do things faster than we would have going after 'doable' goals, and we have enough confidence now to set new stretch targets."

Finally, GE pays strong attention to embedding new processes and ways of working within the organization. "You can talk—you can preach . . . about a 'learning organization,' but . . . reinforcing management appraisal and compensation systems are the critical enablers . . . if rhetoric is to become reality."[6]

Too often, organizations that invest strongly in best practice neglect the fact that it is only "best" practice if it leads to "best" performance. Welch maintains an unrelenting pressure on organizations to perform.

THREE TRAJECTORIES FOR IMPROVEMENT

These three case studies illustrate how three organizations developed highly effective improvement trajectories that began with recognizing their starting point (see Figure 12.7). Mercedes started with high performance being achieved despite practice limitations—a vulnerable position often associated with complacency. U.K. High Tech started with a major investment in practice improvement that was not being translated effectively enough into performance. The GE approach has been based on the need to improve both practice and performance.

The Mercedes case study illustrates the challenge for companies in similar situations to realize their vulnerability when they are enjoying

Figure 12.7
Different Trajectories of Improvement

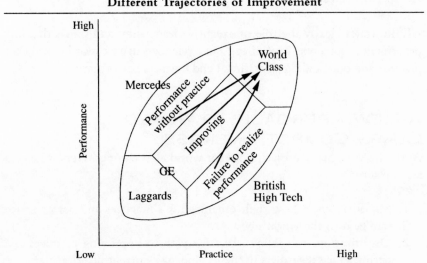

success—often masked by complacency. Although examples may be found in all countries, this is particularly true of companies that have been on top of their industry for years.

Initiative overload, as found at U.K. High Tech, is something to which most managers can relate and, in too many cases, can lead to failure to realize some or all of the desired performance outcomes of improvement activities. In today's rapidly changing environment, companies cannot afford to let up on improvement. There are a number of lessons from how U.K. High Tech addressed these issues. Their actions were characterized by:

- Choice—focusing on a limited set of actions.
- Alignment—ensuring that actions are aligned to external strategy.
- Coherence—ensuring that improvement strategies are coherent rather than unrelated improvement programs.
- Implementation—communicating and managing of implementation throughout the organization.
- Performance orientation—focusing on realizing the performance objectives of improvement programs.
- Perseverance—sustaining an improvement strategy over time.
- Learning—learning internally from the implementation experience and externally through constant monitoring of best practice and performance.

GE illustrates clearly the role of executive leadership; and shows that improvement is not a one-time effort, but a relentless drive toward world class practice and outstanding operational and business performance.

A New Approach to World Class Competitiveness

At the heart of many prescriptions for world class competitiveness lie two assumptions:

1. Managers know how their organization's practices and performance rate next to the world's best.
2. Organizations can follow the same path to improving practices and performance regardless of their company's current position.

Understanding and overcoming complacency is the critical first step in beginning the journey to sustainable competitive advantage. Even companies with high performance may be vulnerable to unforeseen future competition that may mask the Achilles' heel of weak practices.

For all companies, complacent or realistic, understanding the improvement agenda will lead to a clearer improvement strategy. The approach that we have outlined will help executives avoid the "initiative overload" paralysis by offering a discipline for thinking rigorously and systematically about improvement.

NOTES

1. In an interview-based study of 663 manufacturing sites in Finland, Germany, Netherlands, and the United Kingdom, levels of use of leading-edge practices and the resulting performances were measured. "World Class" level of use and practice was defined as a score of 80 percent of both practice and performance scales. For further details, see Hanson, Voss, Blackmon, and Oak, *Made in Europe—a Four Nations Best Practice Study* (London: London Business School, 1994).

2. See P. Ahlstrom, C.A. Voss, and K. Blackmon, "Manufacturing Improvement and Learning—an Empirical Study" (paper presented at POMS Conference, Santa Fe, March 1998).

3. GE Annual Report, 1995.

4. *Business Week,* June 9, 1988, 40–51.

5. GE Annual Report, 1999.

6. GE Annual Report, 1996.

CHAPTER 13

MANUFACTURING IN ITALY: COMPETING IN A DIFFERENT WAY

ALBERTO DE TONI, ROBERTO FILIPPINI,
CIPRIANO FORZA, and ANDREA VINELLI

$\mathbf{M}$ost of our examples of high performance manufacturing (HPM) have come from large plants that use repetitive production processes. Are HPM concepts equally relevant in smaller plants, or is HPM limited to large manufacturers? Will HPM practices apply in a lower volume, less repetitive context? What about the role of strong and ideological labor unions? Is cooperation between labor and management critical for the successful application of HPM principles? Italy presents an ideal case study of the extent to which HPM concepts are universal because some of the structural conditions under which HPM was initially developed are very different from those found within the Italian industrial system.

One of the main distinguishing features of Italian manufacturing is that it consists almost exclusively of small and medium enterprises (SMEs). Analysis of the applicability of the principles of HPM in Italy is particularly interesting, given that these principles were originally developed in the context of large manufacturing facilities. For example, upstream and downstream relations within the supply chain change markedly in SMEs, which have much less bargaining power than large firms. Because of the size of Italian plants, networks of plants are more important than they are in other countries. Thus, study of manufacturing in Italy cannot be restricted to one plant's experience but must necessarily involve the entire network of plants involved in the transformation process.

In addition to size, a further distinguishing feature of Italian plants is that their production is not highly repetitive. Many HPM practices were first developed within the automobile sector, where production systems are highly repetitive. Such systems are typically characterized by many factors, including centralized and bureaucratic management, lower worker-skill levels, higher overhead, low levels of inventory, and low cost of goods sold. In Italy, however, manufacturing plants often produce only in small lots, which makes their systems closer to intermittent (batch) production. Such systems are likely to be characterized by a different set of factors, including decentralized management, higher worker-skill levels, lower overhead, higher levels of inventory, and higher cost of goods sold. Thus, it is interesting to analyze the degree to which the principles of HPM are able to permeate systems with different characteristics from those for which they were developed.

The third aspect that differentiates the Italian industrial system from manufacturing in other countries is that trade unions with a strongly ideological approach to labor questions have always played an important role within Italian plants. This is reflected in national politics and has permeated the entire country, with a high level of conflict in union-management relations that has only recently begun to be defused by political changes in Europe, especially those in Eastern Europe.

This chapter seeks to offer some answers to the following questions: To what extent can HPM be transferred to small and medium firms? Will the use of intermittent manufacturing and the presence of strong unions make the implementation of HPM more difficult? Which methods have been used and what solutions have been found by Italian SMEs when adapting and developing new technological and management practices to their unique characteristics?

THE ROLE OF SMES IN THE ITALIAN MANUFACTURING CONTEXT

There are some important considerations that help in understanding the manufacturing context in Italy. These include the fundamental role played by small and medium plants in Italy and the level of internationalization of production in Italy.

Analysis of the top 5,000 Italian firms (in terms of sales) helps in identifying the main characteristics that distinguish Italian industry. First, there are very few large firms; only the top 40 percent or so have sales of more than 1,000 billion Lira (about $50 million). Second, about 1,500 SMEs fall into the band of firms with sales of from 100 to 1,000 billion Lira (about $50 million). Such firms are, effectively, the backbone of the Italian economy.

The Italian economy can best be understood if the Italian situation is compared with that of the most industrialized European countries. In 1992, Italy had the largest number of firms of any country in Europe. It had 3,250,000 firms, 21.1 percent of the entire nonagricultural private sector in the EU. Italy was also the country that had the smallest average-size firm (see Table 13.1). The total number of firms in Italy equals the number of firms in Spain, Austria, Greece, and Ireland combined. Italy, France, Germany, and the United Kingdom, among them, account for two-thirds of all manufacturing production in the EU.

The small firm, especially one that is still at an early stage of its cycle of development, is often still run by the entrepreneur-founder. Together with the family firm, the small firm is one of the basic social units in Italy, a country that places a high priority on individual and family values, often at the cost of social and collective values.

The reason why Italian SMEs rarely grow much bigger is the subject of a great deal of debate in Italy. Some argue that the small size of Italian firms is linked to limitations within the entrepreneurial and managerial culture of the country and that this limited development is due to limits that are inherent within the model of a firm that has been adopted. Others have highlighted structural reasons for nongrowth, such as market opportunities, the role of state participation, and the role of national financial markets.

All conditions being equal, the size of a firm is linked to the size of its "natural" market in terms of geography, history, and linguistics. The natural market in Italy is much smaller than that of nations with larger populations, such as the United States or Japan; or nations that have large areas of political and economic (ex-colonial) influence, such as the United Kingdom; or nations where the same language is spoken in other nearby countries, such as Germany. These factors all inevitably expand the "catchment" area of firms' markets.

Furthermore, until quite recently, state-owned firms were almost monopolists in several key sectors in Italy. These sectors consequently benefited greatly from economies of scale and were traditionally led by very large firms. Key sectors in which state ownership dominated included electricity and gas, iron and steel, telecommunications, petroleum, chemicals, railways, transport, infrastructure (roads, etc.), defense, heavy industry, and shipbuilding. In these sectors, as well as others, competition was limited, or even abolished, by law, to the advantage of the state-owned firms. Today, the situation is changing as the Italian state has begun a process of privatization in many of these sectors.

Table 13.1

Gross Domestic Product (GDP) per Inhabitant and Size of Firm in the EU in 1992

	Italy	France	Germany	Spain	United Kingdom	Total EU
GDP per capita (ECU) Percent	16,469 104.5%	17,676 112.2%	17,080 108.4%	12,161 77.2%	15,448 98.1%	15,754 100%
Number of firms Percent	3,250,000 21.1%	1,960,000 12.7%	2,425,000 15.7%	2,170,000 14.1%	2,525,000 16.4%	15,425,000 100%
Average size (Employees per firm)	4	7	10	5	8	7

Source: Istituto Tagliacarne, 1993.

The Italian financial market is unusual. Very few firms are quoted on the stock market, where new capital for development is raised. In addition, the credit banking system has always tended to direct capital to certain specific sectors or areas of the country.

However, many Italian SMEs have traditionally been quite competitive. The factors on which the competitiveness of Italian SMEs is based include:

- The ability to develop the innovations demanded by the market.
- Flexibility and adaptability to market and environmental conditions.
- A strong sense of unity at the management level and close identification with the entrepreneurial and managerial aspects of the plant.
- Good possibilities for self-financing, due to a long-term tendency to reinvest profits in improvements.
- The ability to create a united atmosphere within the plant, often better than that found in large companies.

Particularly important are "design" abilities of Italian firms, including the ability to define a wide product range and the ability to customize products. A large number of small Italian plants have excelled in their ability to develop and implement innovations that have been well received by their customers and that have remained protected from potential new competitors by a sort of invisible barrier, due to the image that products made in Italy are often better than those from abroad.

However, some weak points have persisted in Italian manufacturing, including the lack of ability to bargain in the resource market (especially financial and technological), low propensity to form "critical masses" to confront institutions (through the setting up of trade associations) or markets (through joint ventures, associations, etc.), and the general inadequacy and inefficiency of public and private infrastructure (transport, roads, railways), all of which are elements that particularly affect smaller plants.

THE INTERNATIONALIZATION OF ITALIAN INDUSTRY

Over the past 10 years, Italian industry has undergone a marked process of internationalization. As recently as the mid-1980s, Italian investment abroad was still very modest. Today, however, the total of Italian investment outside Italy matches the total of foreign investments within Italian industry. This recent international expansion has been led by small and medium manufacturers. Over the past decade, more than 9,000 Italian firms have experimented with different forms of internationalization (see Table 13.2). The current situation is the result of 10 years of important changes that encouraged the integration of Italian industry within world markets. Even though the amount of foreign investment in Italy is still slightly higher than the amount of Italian investment abroad, it is clear that Italian industry is undergoing a process of internationalization. Two main factors have contributed to this process: (1) the initiative of groups of medium-sized plants active in the traditionally competitive sectors of Italian industry and (2) the growing involvement of both small and medium plants in growth processes abroad.

The distinguishing feature of the Italian presence at the international level during the early 1990s was the vast increase in the number of Italian firms involved. This was due to the increased involvement of small and medium plants in the internationalization of the production process. Of the 622 Italian multinationals, 350 (56.3 percent) have less than 200 employees, and 478 (76.8 percent) have less than 500 employees.

THE ADOPTION OF HIGH PERFORMANCE MANUFACTURING IN ITALY

The process of internationalization should be viewed within a context of much wider change in the competitive environment, where Italian plants have operated in a dynamic manner. The level and the type of performance required to be internationally competitive is rising constantly and affects all types of performance: costs, productivity, internal quality, quality vis à vis customers, time required to introduce new products, and delivery times.

Table 13.2
The Internationalization of Italian Industry in 1995

	Direct Italian Investment Abroad (A)	Direct Foreign Investment in Italy (B)	(B/A)
Investors	622	966	1.58
Firms involved	1,842	1,630	0.88
Employees	595,547	527,461	0.89
Turnover (in thousand millions Lire)	156,841	212,175	1.35

Source: Cominotti, Mariotti, 1996.

For SMEs, their new competitive situation is mainly the result of three factors working simultaneously. First, globalization of the market has increased both the quantity and the quality of the plants against which Italian plants must compete. Customer needs and requirements have also tended to expand, corresponding to improvements in supply. Second, internationalization of the production chain has occurred, based on where the phases of the production cycles are located. This has occurred as firms have searched for both lower costs of production and certain skills required for each phase of the cycle. This has had the effect of raising the performance levels below which a plant will not be competitive on costs and quality. Third, HPM has led to important changes in the performance of Italian plants, allowing them to attain levels of excellence on several dimensions simultaneously.

Furthermore, it should be noted that Italy's traditional competitive factors of recent years have become less and less important. In particular, the economic unification of Europe and the introduction of a single European currency, the Euro, has rendered redundant the development policies based on export to countries with strong currencies (e.g., Italy to Germany). Thus, factors concerning inefficiency, rather than competitiveness, have come to light. Such factors would formerly have been masked by favorable currency exchange situations. These factors have forced Italian plants to set out along the path to renewal. Which paths have Italian plants chosen, especially those plants that compete in international markets? Which production policies and strategies have they chosen? Are Italian SMEs trying to implement HPM *tout court*, based on the overseas literature and practices? Or has there actually been an Italian HPM approach, with adapted practices?

We offer some answers, based on our comparison of the best practices of high performing plants in the Italian sample with those of the international sample. In the following sections, we will try to highlight the various aspects of adoption and adaptation of HPM practices as carried out by Italian plants. We will identify specific factors in order to be able to offer some generalizations regarding the applicability of HPM within the Italian manufacturing context.

Best Practices: Comparison of High- and Low-Performing Plants

We studied two separate groups of Italian plants, the high and the low performers, identified on the basis of their performance. The ways in which the two groups adopted innovative practices was analyzed by differences in production processes and production technology, management of production flows, quality management, information systems, human resources, and organization and production strategies. The two groups exhibited statistically different behavior in many of the specific areas analyzed. There were significant differences between the two groups in 83 percent of the practices.

The practices where the differences between high-performing and low-performing plants were most marked are shown in Figure 13.1. There is a marked difference between the two groups in all areas of operations. This confirms the innovative nature of HPM and the impact of these new approaches on performance. HPM is not characterized by the use of unique and exclusive technologies, but rather by the level of integration and the coherence of *all* management and organizational innovations that affect and alter the traditional principles of both organization and management.

Best Practices: An Intercountry Comparison

Comparison of Italian plants with those of the other countries involved in this research revealed considerable differences in the way in which high performance practices are applied. Figures 13.2 and 13.3 show the practices adopted in Italian plants that differentiate these plants from those in other countries. Figure 13.2 shows the differences for the high-performing plants and Figure 13.3 shows them for the low-performing ones.

For the high-performing plants, 40 percent of the variables considered showed a significant difference when Italian plants were compared with plants from other countries. The variables that concerned relations with customers (see Figure 13.2) were, on average, better in the high-performing Italian plants than those in high-performing plants in other countries. Customer satisfaction, involvement in quality, and information concerning products received by customers are the variables that differ most markedly in favor of the high-performing Italian plants. This highlights a basic strength of Italian plants, something that emerged during our visits to plants and from discussions with various managers. Italian plants have the ability to respond to market demands to be flexible and to customize products. The product is considered to be the fundamental point of contact between the plant and its customer; thus, a great deal of attention is paid to it and to the plant's ability to customize and to personalize what is produced.

This comparison also revealed the importance that Italian plants place on employment stability. In all plants (both high and low performers) the workforce was more stable than in the other countries studied. This is a

Figure 13.1
Comparison between High- and Low-Performing Plants in Italy

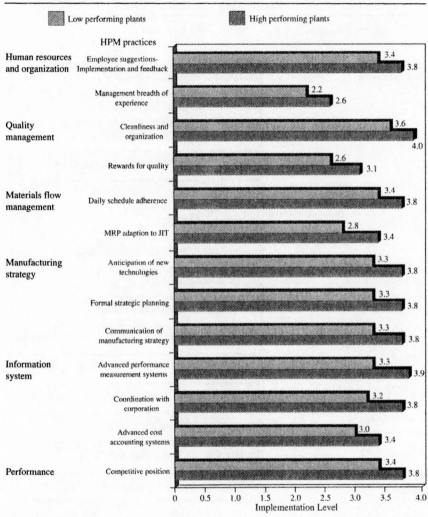

Source: HPM data.

result as much of the laws and norms that protect employees in Italy as of a conscious policy choice on the part of the plants themselves.

Quality management is viewed differently in Italy than in Japan or the United States. Italian plants envisage the idea of quality management as the search for customer satisfaction and customer involvement in the quality of

Figure 13.2
Comparison between High-Performing Italian Plants and
High-Performing Non-Italian Plants

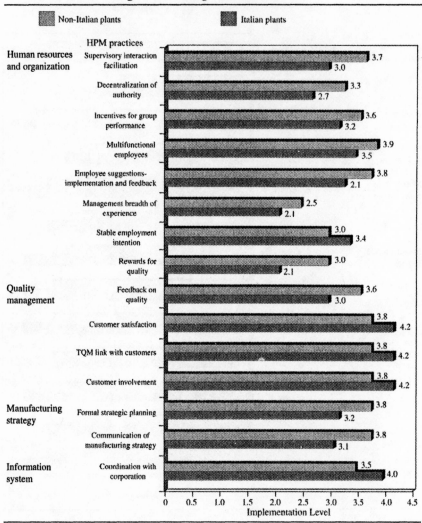

Source: HPM data.

products. Other methods, such as using work groups and giving greater responsibility for their work to the employees themselves, have not been used very much as tactics for increasing employee involvement for continuous improvement. Thus, Italian plants typically do not adopt all the various quality management approaches as they would if they were to follow

Figure 13.3
Comparison between Low-Performing Italian Plants and
Low-Performing Non-Italian Plants

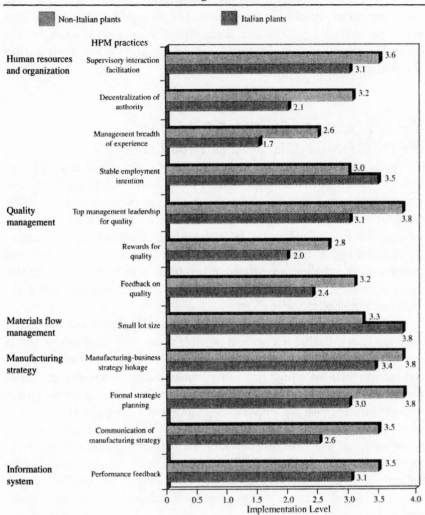

Source: HPM data.

the "total approach to quality" inherent in the concepts of HPM. However, it should be stressed that in recent years many Italian plants have begun to make considerable efforts to adopt more comprehensive quality systems, such as ISO 9000, and, more generally, to adopt a *total* quality management (TQM) approach.

Italian plants lag farthest behind in the area of management. There is a lack of training for management in the skills required for HPM. This problem is both illustrated and accentuated by the fact that traditional entrepreneurial styles of management still prevail in Italy. Few Italian plants periodically reformulate their plans for long-term strategy, nor do they communicate their strategy to personnel at other levels within the plant.

THE ITALIAN PATH TO HIGH PERFORMANCE MANUFACTURING

The path taken by the Italian high performers is thus somewhat different than the path followed by high-volume plants with mass production. Italian plants have not introduced flexibility starting from repetitive production; rather, they have sought to achieve efficiency, starting from highly intermittent batch production, by maintaining or developing their capacity to make highly customized products, further opening up markets and customers. The challenge for Italian plants today is in knowing how to improve efficiency, delivery times, and quality without penalizing their flexibility of response to the market.

Study of the experiences of the Italian plants that have started to introduce HPM approaches has made it possible to identify some salient points regarding the process. In terms of the goal of improving efficiency levels and linking them with high flexibility, Italian plants have tended to choose those paths that make personalized response to the market possible. Italian high performers have not emphasized adoption of a complete HPM model, intrinsically complete and structured. Rather, they have adopted concepts, techniques, and practices that will serve to guarantee personalization (customization) of the product as a flexible response to the market and gradual approaches to organizational change, proceeding by accumulation rather than by "jumps." Thus, the objectives most commonly adopted by innovative Italian plants are:

- Coordination and integration among the supply, design, industrialization, production, and distribution phases.
- Actions to improve the quality of products and services for customers.
- The will to change and innovate the attitudes and the policies of human resources.

The internal (plants') and external (context) variables that seem to influence positively or negatively the adoption of best practices in small and medium plants are listed in Table 13.3.

The picture that emerges from our discussions with managers studied is that of a culture of young, dynamic plants, with a high propensity to exploit situations of competitive advantage from time to time, rather than to develop elaborate long-term market strategies. Product innovation is a fundamental competitive factor, but it is often not linked to marketing activities and is not planned in the medium term.

Importance of Performance for Italian SME High Performers

The types of performance that successful Italian SMEs see as being most crucial involve their ability to satisfy customer requirements in a personalized

Table 13.3
Opportunities and Potential Problems for Italian SMEs Wishing to Introduce High Performance Manufacturing Practices

	Opportunities	*Potential Problems*
Firm variables	Good ability to perceive and respond quickly to customer requirements.	Broad-range and low-unit volumes, difficulty of medium- and long-term planning.
	Low level of conflict in SMEs	Lack of experience in competing in highly globalized markets (often niche markets prevail).
		Scarcity of managerial skills (mainly linked to the size of the firm).
		Lack of know-how about managerial aspects (e.g., modeling and formalizing) and on technical-specific aspects (e.g., marketing, design techniques).
Variables in each context	General willingness to change typical of an industrial context characterized by new, young SMEs operating throughout the country.	Rigidity of the norms that regulate the labor market.
		Low level of bargaining power when dealing with suppliers and customers.

manner, at lower costs and with the desired level of quality, in terms of innovation and customization of products, product costs and services offered to customers.

- *Time.* Time is considered to be a critical performance variable, even though Italian plants do not typically make large investments explicitly designed to improve the speed of response to the market, delivery times, and the ability to reduce the effects of market variations.
- *Quality.* Consistent quality is seen as being very important by Italian plants. All too often, however, Italian plants pursue quality solely through ISO 9000 certification, which is perceived more as an end in itself than as part of a continuous process of improvement. Some plants view quality as a cost that could lead to greater rigidity. Nevertheless, they do perceive the aims of quality in a positive light, a goal that means not only the lack of defects but also increased satisfaction of customers' requirements.
- *Costs.* In the past, cost was not considered to be a particularly important variable in Italy because the lira has always tended to devalue against other currencies. Thus, plants did not need to become cost sensitive. With the advent of the Euro and a stable exchange rate, costs and efficiency have become much more important to Italian plants. Cost control is, however, a problem that has yet to be taken up in a serious manner, and costs do not always play a large enough part in decisions made by Italian plants.

Importance of Best Practices by Italian SME High Performers

- *Product Standardization.* This is still an underused concept in the design phase in Italian plants, most likely a consequence of the type of plant we are studying (wide range of products and low-unit volumes). In particular, there appears to be little use of either "design for manufacturing" techniques or of standardization of models and methodologies in Italian plants. There is, however, a tendency to standardize components. There is also a tendency to move personalization of products farther and farther downstream and to set up multifunctional design groups.
- *Total Quality.* Italian plants are very sensitive on this issue and are willing to invest in it. However, it should be stressed that, in reality, only specific techniques drawn from TQM practices tend to be adopted and applied, rather than a more global and complete approach to total quality.

• *Production Processes.* There have been many initiatives made in terms of process improvement in Italian plants, among which are focusing of processes (specialization and externalization of phases) and improvements in process technologies that aim to improve efficiency, to reduce throughput times, and to increase the flexibility of the product mix through rationalization of layout, the use of proprietary equipment, and reductions in tooling time.

• *Just-in-Time.* JIT is rarely used in Italy. Indeed, in intermittent production (typical of SMEs with many products), flow time cannot be reduced beyond a certain limit. Furthermore, the supply network of small and medium plants has specific characteristics that make the use of JIT techniques both difficult and not really worthwhile.

• *Information Systems.* Information systems are widely used in Italy and often have tailor-made solutions for design and production planning. They are less commonly used within the production process and for statistical checks on processes.

• *Human Resources.* Flexibility of labor is a strong point in Italy. It is essential for flexibility, especially volume flexibility. The plant's own culture and that of its employees (willingness to change) offer an ideal opportunity for the introduction of specific HPM practices.

Reduction in the number of levels in a plant's hierarchy and the cutting back of organizational structures has not been a major initiative in Italy—the staffs of small and medium plants have always been "lean." A lean management structure reinforces direct dialogue with employees and favors greater responsibility and multifunctionality of those who work in the plant. If there is no plan for coherent organization or for decentralization of the decision-making process, there may be evolution "at the top," which encourages a paternalistic or family style of management. This tends to strangle innovative concepts with traditional management limiting the possibility that they will have a positive effect.

Many plants see the professional and cultural growth of human resources as a strategic objective. But Italian SMEs often find it difficult to carry out efficacious human resource development policies because they lack the adequate structures and skills required (e.g., support for developing incentive systems for personnel, structures for carrying out training, development programs aimed at meeting specific needs, etc.).

• *Suppliers.* The Italian supply system is made up of two distinct types of suppliers: (1) prime suppliers, which have a good bargaining position, and

(2) small scale suppliers, or artisans, with low bargaining power and high levels of flexibility.

Supply-chain management approaches typically seek to ensure reliability, punctuality, and quick response times to variations in quantity and mix on the part of the suppliers. The most widespread interventions in Italy concern helping small suppliers to grow and identifying the first level of suppliers. Trends within the supplier management process include adopting systems for evaluating and selecting suppliers and setting up medium- to long-term relationships with them. Some HPM practices are not adopted because they are either not feasible or not practical. There may be a strong comakership relation with first-level suppliers and JIT within the supply function, that is, frequent deliveries of small lots and large reductions in the number of suppliers. On the other hand, some SMEs that supply large-scale high-performing plants do take part in more HPM initiatives.

• *Customers.* Downstream production relationships are considered to be very important, if not decisive, in Italy. Italian SMEs pay a great deal of attention to customer satisfaction, to customer involvement in defining quality, and to using information gathered from customers in order to improve product quality. The most widely used HPM techniques by Italian manufacturers are increased in-depth knowledge of customer requirements, shortening distribution channels, adoption of dedicated information systems, and developing and increasing the skills of personnel.

Human Resources and Industrial Relations

In terms of human resource management, we learned several interesting things during our discussions with Italian managers and entrepreneurs, which serve to outline the objectives that the high-performing Italian plants are trying to achieve. The cultural growth of a management culture is a necessary but not a sufficient condition. "The new" must also be passed on, which must necessarily involve the entire organizational system. Individual skills must be transferred and translated into procedures, behavior, and style of management. Greater responsibility of personnel, participation, and multifunctionality all require a change in the way that the organization of work is viewed and in the systems through which work is assigned and rewarded.

Work groups set up in order to improve quality, efficiency, and service offer a valid tool for solving many production and management problems. However, work groups should be set up using suitable approaches. The cultural and professional growth of employees should be sought continuously. There must be improvement in the quality of industrial relations. Lack of knowledge regarding high performance production can lead to adverse reactions from trade unions and could potentially block the development process.

One of the most important developments that has taken place recently in Italy has been the launch of a process of developing responsibility in employees. This has affected all levels of personnel, from the top to the bottom. This "revolution" in the employee's role implies that there has been a rethinking of the role of the supervisors and of the role of the trade unions. Even though this is never a simple process, SMEs do have some advantages over large plants in this regard. There are already far fewer levels in the hierarchy, and workers are less likely to be strongly unionized. Where it has been tried, the process of expanding the role of all employees has been met with an excellent response, which serves to demonstrate how human resources are as yet untapped in many Italian SMEs.

The experience of on-the-job training by expert colleagues, learning through training courses on product and process technologies, setting up work groups with collective rewards, and delegating have all shown how knowledge can be copied and reproduced within a plant. This has had a marked impact on the level of involvement, willingness to participate in problem solving, and obtaining small but cumulative results, which are part of the philosophy of continuous improvement.

A number of Italian plants are also experimenting with incentive schemes, such as rewards for all workers linked to the gross operating margin of the plant. These experiences not only act as an incentive for productivity, but also have shown how they help orient workers in the direction of the plant's aims and objectives. Overall, the Italian unions' response to such changes has been positive; and plants, which historically were always a place of division of labor, are being transformed to focus on personal and professional growth. In those plants where initial diffidence has been overcome, the unions are now actively participating in the reorganization of production methods and in formulating contracts for "rewards for results." Within this phase of development of industrial relations, a new "bottom up" approach has been tried, rather than the more usual "top down" approaches. The most successful initiatives have almost always been initiated by plants and local workers' representatives, rather than by employers' associations or the national committees of trade unions.

SOME MANAGERIAL IMPLICATIONS OF THE HPM APPROACH IN ITALY

Several points have emerged from the meetings and discussions we held with managers of the Italian plants studied regarding the application of HPM in Italy. HPM models are based on a new philosophy; thus at least four major changes must be carried out in the way in which an organization is structured:

1. Changes in "internal" relations (involvement of workers)
2. Changes in "external" relations (with customers and suppliers)
3. Changes in flow management (production flows become leaner, sources of interruptions and variations are eliminated, etc.)
4. Changes in the search for total quality within the plant

HPM requires a long-term effort that may last for years, one that entails both small and large interventions in all areas of the plant, not only in the area of production. In Italy, many often radical adaptations and changes have had to be made to HPM in order to take the specific nature of Italian plants into account. Some practices, for example, those involving suppliers or flow management, cannot be "slavishly" transferred to the Italian SMEs.

There are two main obstacles to implementing HPM in Italy: (1) Low product-unit volumes are exacerbated by insufficient standardization during the design of components, products, methods, and concepts. (2) There is a lack of appropriate skills required for the new methods, a problem that is accentuated by the traditional style of management of Italian plants. However, the concepts and the methods HPM is based on are valid and can be adapted because they are oriented toward a mix of excellent performance that SMEs may already be able to achieve; they render many improvement interventions that SMEs are already involved in synergistic, and they help integrate them into an overall, coherent logic.

The passage from the logic of adapting individual techniques and methods to an HPM strategy, where both internal and external relations change radically, requires a change in the plant's perspective. Plants must move from single interventions, often with short-term aims, to integrated medium-term policies. This is the most difficult, but also the most promising, change necessary. In order to control costs, advanced systems of measuring performance and tools for evaluating the overall return on investment in technological-organizational innovations are required. The quality of the product or service should not stop at the standards set by ISO 9000 certification but must be interpreted as the result of a continuous process of improvement toward meeting the customers' requirements.

Product innovation is still a winner in Italy, but it must be more closely linked to marketing strategies and to technological innovation. Knowing one's own customers well is not the same as knowing the market well: both the knowledge and the professionalism of Italian management must be developed in the areas of marketing and strategic planning.

Personalization of products is already well developed in Italy. It must be matched by design efforts (modularization, standardization, etc.), so as not to increase the internal variety of products, which is one of the main sources of increased costs. In addition to standardization of parts and product modularization, a higher level of standardization of production processes is also considered to be an important goal. It should start from a careful analysis of the reasons behind any obstacles, such as management's lack of know-how or the development of suitable and inexpensive technologies.

Single sourcing for each part number is not a viable alternative for SMEs for a variety of reasons, mainly related to the need for flexibility of volume. However, it can be used as an underlying principle: that of setting up long-term relations with a small number of suppliers. New approaches to designing products can make it possible to increase the volumes of each part number bought, thus increasing the plant's bargaining power in relation to its more important, or first-level, suppliers and allowing smaller suppliers the opportunity to grow. Management and growth of a network of small suppliers, which guarantee lower costs and good service, is seen as being important. However, equally important is the idea of rationalization of supply chain relationships (first level of suppliers, evaluation/selection systems, and a tendency to develop medium- to long-term relationships). Many other support methods should also be strengthened in Italian plants, in particular, information systems for programming production and process control and for managing a plant's upstream (suppliers) and downstream (customers) flows.

In the final analysis, we could conclude that in their march toward the innovation and adoption of high performance models, Italian SMEs intend to maintain their flexibility of response to the market at a high level, in terms of offering a wide range of products and customization of products, with the aim of reducing costs. They believe in the uniqueness of Italian entrepreneurship and are willing to gradually change and adapt others' models (both of product and of process) only if they are not in conflict with the plant's current strategies. Italian plants seek to constantly improve their relations with customers. They tend to integrate operating phases, upstream and downstream, utilizing the abilities and the specialization of other plants within a network logic.

REFERENCES

Bagnasco, A. *La costruzione sociale del mercato. Studi sullo sviluppo di piccole imprese in Italia.* Il Mulino, Bologna, 1988.

Beccattini, G. "Dall'impresa alla quasi comunità: dubbi e domande." *Rivista di Economia e Politica Industriale,* n. 68, 1990.

Cominotti, R., and S. Mariotti. *L'Italia multinazionale.* ETASLIBRI, 1996.

EU. Official Journal of European Communities, nr. L 106/6, Bruxelles, 1996.

ISTAT. *Conti economici nazionali,* Roma, 1998.

Istituto Tagliacarne. *Conti economici,* Roma, 1993.

CHAPTER 14

GERMANY: PURSUING THE
TECHNOLOGY PATH

PETER MILLING, KATHRIN TUERK, and SVEN WEISSMANN

$\mathbf{M}$ost German manufacturing organizations have achieved competitive success through the use of hard technologies. In their use of this strategy, they have done extremely well, achieving the highest average productivity levels in our global sample. This strategy has been very focused in German organizations, which have been much slower to pursue soft technologies. For example, many German organizations have begun to implement supplier management, simultaneous engineering, and quality management initiatives only recently. Why have German plants evolved in this manner? We would argue that examination of the strategy of German manufacturers provides an excellent case study in capitalizing on the unique strengths of a country's environment.

IMPORTANCE OF MANUFACTURING
IN GERMANY

Manufacturing has always been very important in Germany, and its international competitive advantage has been influenced by both national and organizational factors. Some factors, such as geographic location, educational level of the workforce, political stability, and demand conditions, affect every organization within the country. Other factors are more unique to specific organizations, resulting from local (organization-specific) management initiatives. In this comprehensive analysis of the global success of German manufacturing, we consider both national and local factors.

Like many developed countries, Germany is experiencing a transition from a manufacturing economy to a service economy. Since the beginning of the 1970s, the number of persons employed in the manufacturing sector in the Federal Republic of Germany has decreased continuously. Whereas in 1970 almost half of all Germans were employed in manufacturing, this had dropped to one-third by 1997. During the same period, the fraction of the employees in the service sector increased from 43 percent to 64 percent. This transition mirrors the changes experienced in other developed countries during this same time period. In the United States, for example, this evolution was even more extreme. In 1950, half of all Americans worked in the manufacturing sector; it is currently only 20 percent. An employment rate of more than 70 percent in the service industry shows the dominance of services in the United States, compared with other industrialized nations.

In recent years, practitioners and theoreticians have referred more and more to the strategic role of manufacturing in the economy. They emphasize that efficiency, effectiveness, quality, and flexibility of production are the keys to success in the competitive environment. Just producing "good" products is no guarantee of long-term success. The procedures and the processes used to make products are more important. There is an increasing influence of manufacturing on the competitiveness of industrial companies. Hence, the manufacturing function should be integrated into corporate strategic planning and in the formulation of competitive strategies.

To an increasing extent, competition has been taking place between companies operating in the triad (Japan, the United States, and Europe). Often, Europe is seen as being in a favorable position because of the size of its market. However, there are differing opinions concerning the relative position of Germany. On the one hand, Germany is considered an exceptionally strong competitor; on the other hand, there is discussion of an endangered position of Germany in the competitive environment. German companies are confronted with powerful and service-oriented competitors from countries with much lower cost levels, both in Germany and in international markets. As competitors have improved their operations and efficiency, pricing has become a decisive factor in recent years. Over the years, German industrial firms have been perceived as high-priced but also high-quality manufacturers with reliable delivery dates. This is consistent with their focus on a hard technology strategy. Historically, there were good reasons for the German companies to follow such a manufacturing strategy. A brief comparison between Germany and other countries explains why.

GERMAN COMPETITIVENESS

How do German plants compare with top-performing plants internationally? In order to examine this question, we selected a number of important measures of performance from the high performance manufacturing (HPM) database, shown in Figure 14.1. We then plotted the average performance from the country in our sample that had the best performance on that measure and compared it with the average performance of the German plants we studied on the same measure. In order to make comparisons across performance measures more easily, we standardized each of the measures. Thus, a value of 0 indicates performance at the average for the five countries we examined. Similarly, a positive value indicates performance that was better than average, and a negative value indicates performance that was worse than average.

We found that German plants were particularly strong in cost and flexibility. German plants were also above average in quality and inventory turnover, ranked second to the United Kingdom. However, we found that German plants were less competitive in delivery time and on-time delivery, where they ranked behind the Japanese and the U.S. plants. Thus, German plants appear to compete on flexibility and cost, relative to their global

Figure 14.1
Performance Comparison of Plants

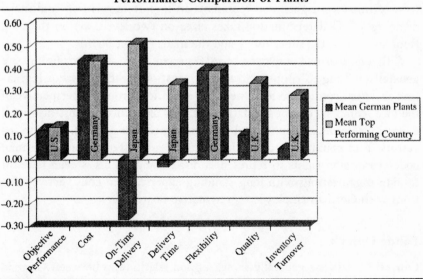

Source: HPM data.

competition. This is consistent with their emphasis on hard technology as a path to competitive success.

International competitiveness is determined by many different factors. An organization exists in a particular environment that provides it with basic factors, including natural resources, climate, and geographical position, as well as workforce and legal and economic conditions. In addition to these basic factors, more specialized factors are developed in a country, including communication technology, technical infrastructure, and the development of highly qualified workers and research institutes. In the following sections, we will examine the factors that contribute to the success of German plants in global markets.

FACTORS CONTRIBUTING TO GERMAN COMPETITIVENESS

Legal System

An important environmental factor in Germany is its legal system. In particular, codetermination rights and labor legislation provide an environment that is different from many other countries. Worker participation in management decisions is *required by law* in German private industry (codetermination rights) and represents an important form of "industrial democracy." Codetermination takes place on two levels: (1) on the shop floor and (2) on the supervisory and the management boards.

Although German employee board representatives and labor unions are generally willing to cooperate, codetermination rights and labor legislation are often considered to be negative conditions from the perspective of the local company. Labor legislation includes protection against dismissal, educational leave, and continued payments in case of sickness. Foreign investors from countries such as the United States often consider German codetermination rights to represent a disadvantage. This is even true for foreign organizations with long-standing experience in constructive relations with German employee representative boards.

Labor Unions

Our HPM database generally shows a good relationship between German companies and labor unions. Labor unions are represented in 82 percent of the German plants in our sample. Within these plants, the percent of employees organized in labor unions ranges from 5 percent up to 92 percent, with an average of a little less than 50 percent. Most German labor unions are relatively large, so that only one or two unions are typically represented in a plant. In 60 percent of the plants where unions are active, there is a good relationship between the plants and the unions, with this relationship considered to be very good in an additional 36 percent of the plants (see Figure 14.2). Thus, we found that union relationships were considered to be at least good in 96 percent of the German plants we studied.

German plants have a great deal of flexibility in making special arrangements, for example, plant agreements between the employer and the works council. Liberalization of labor law and Industrial Constitution Law may overcome mobility hindrances and adjustment hindrances on the employee side.

Geographic Situation

Germany's climate and infrastructure support international competitiveness. Since the opening of the boundaries of Eastern Europe, Germany has been located in the center of Europe and, therefore, has an excellent location in respect to transport facilities. Germany has an outstanding road network, as well as a railroad system and sea routes.

Figure 14.2
Relationship between Companies and Labor Unions in Germany

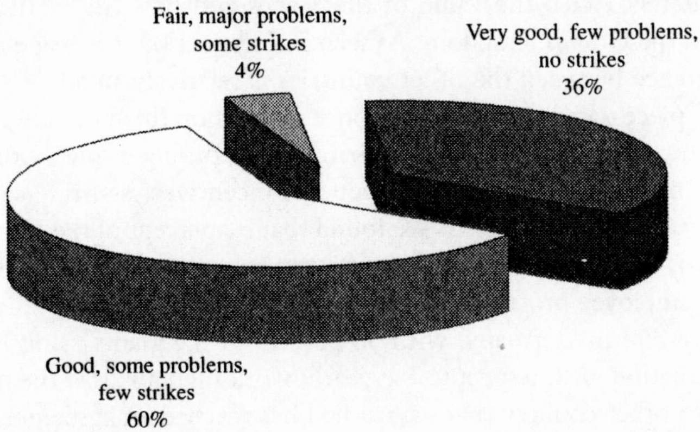

Fair, major problems, some strikes 4%

Very good, few problems, no strikes 36%

Good, some problems, few strikes 60%

Source: HPM data.

With the opening of its eastern boundaries, Germany's advantage and most important feature in international competition within Europe has become its unique position in Central Europe. Because of this, Germany has become a location for European trading centers and business logistics headquarters. The transit function, limited in postwartime to north-south relations, has been expanded to include east-west relations since the beginning of the 1990s. Thus, the economic benefits resulting from Germany's geographic location are far reaching; liberalized markets, harmonized patterns of competition, and an extension of the European Union (EU) to the Eastern European countries are the most important changes. In order to take advantage of the economic advantages resulting from its location and function as a European hub for market and commerce, Germany needs to continuously enlarge its infrastructure strengths, particularly its transportation and communications infrastructure. Furthermore, with regard to the national economies of its East European neighbors, the special location advantages of the eastern states of Germany and of Berlin, in particular, have to be built up through infrastructure, institutions, and political support.

Labor Cost and Payment Systems

German labor costs are high. This is particularly true when we focus on the countries in our investigation, which are also Germany's main competitors in the global market.[1] In addition, German employees receive substantial benefits, including six weeks or more of vacation, special vacation bonuses, social benefits, sickness payments, and Christmas bonuses. This adds up to approximately twice the value of the wages and benefits in the United States and the United Kingdom. As seen in Figure 14.3, the wage and benefit difference between the other countries is relatively small.

While piece-rate payments were once a common form of compensation in Germany, only 23 percent of the firms pay according to piecework today. In German plants, other types of wage and incentive systems have become more prevalent in recent years. We found that 52 percent of the German organizations used group-oriented incentive systems, second only to Japan. In contrast, employee profit sharing at the management level has only a small acceptance rate in Germany, with 30 percent of the plants using it.

The situation with respect to the yearly working hours is at the other extreme. No other country that we studied has reached an agreement in collective bargaining to work fewer hours than German workers. German workers work an average of only 1,579 hours per year, compared with the

Figure 14.3
Average Labor Costs

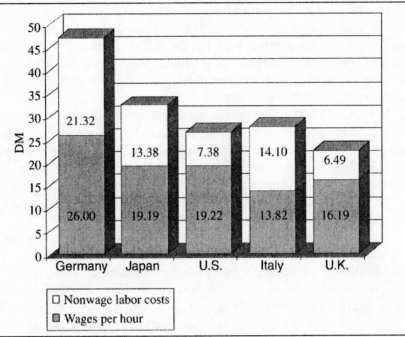

Compare *Institut der deutschen Wirtschaft Koln* (ed.): 1998—Zahlen zur wirtschaftlichen Entwicklung der Bundesrepublik Deutschland, Koln 1998.

Japanese and the U.S. averages of 1,848 and 1,912 hours per year, respectively. Furthermore, rigid German regulations hinder flexible adjustments to market requirements, and machine time per week is also very low.

Worker Skill Levels

Worker skill levels are partly determined by general and professional training and the educational system of a country. Increasingly, German companies criticize the entrance level of German students to universities as too high, demanding a reduction of elementary and secondary school from 13 to 12 years. In addition, the modernization of education has promoted technical knowledge as well as foreign language knowledge. Technical knowledge is traditionally seen as a competitive advantage in Germany. Workers in Germany go through an elaborate vocational training system. They are usually well trained and master a broad variety of skills, making them well suited to work for organizations that pursue a hard technology strategy.

In 1997, 1.8 million students were enrolled in German universities. There were a thousand fewer persons enrolled in the engineering sciences in 1997 than in the previous year, and the declining trend since the beginning of the 1990s continues. For example, 62,600 persons studied engineering sciences in 1992, whereas only 45,500 students were enrolled in engineering in 1997, a total decline of 27 percent.

In 1996, 237,000 students graduated from German universities and technical colleges. More than half of the graduates had university diplomas or even doctoral degrees from German universities. Because of their focus on technical training and education, which includes long periods of vocational training, German students tend to be older when they graduate. On the average, when finishing their studies, German students are 28 years old and have studied a total of 5.7 years at a university.

Many of the employees in the German plants have a technical degree or apprenticeship training. According to the opinion of many managers, the qualification of employees, combined with the professional education system, is one of the prime advantages of Germany. The German labor market is characterized by a high percentage of workers who have served a formal apprenticeship in a particular occupation or trade and have passed a qualifying examination. This is well suited to Germany's focus on hard technology.

Consistent with its lighter emphasis on soft technologies, only 43 percent of the workers in Germany are trained for several activities, in contrast to the United States, where 68 percent of the workforce is cross trained. This ability has increased in importance, however, as soft technologies have gained in importance in Germany. For example, the electric appliance manufacturer Bauknecht, the second-ranked plant in the worldwide sample, has achieved high flexibility, with 80 percent of its workers cross trained.

In terms of productivity, Germany is the top performer among the countries we examined. The other countries in the investigation reached only 55 percent to 83 percent of German productivity (Figure 14.4).

Research and Development

German managers view manufacturing technology as their strength. However, they do not see this technological strength as relevant to high-tech

Figure 14.4
Productivity and Labor Cost per Unit

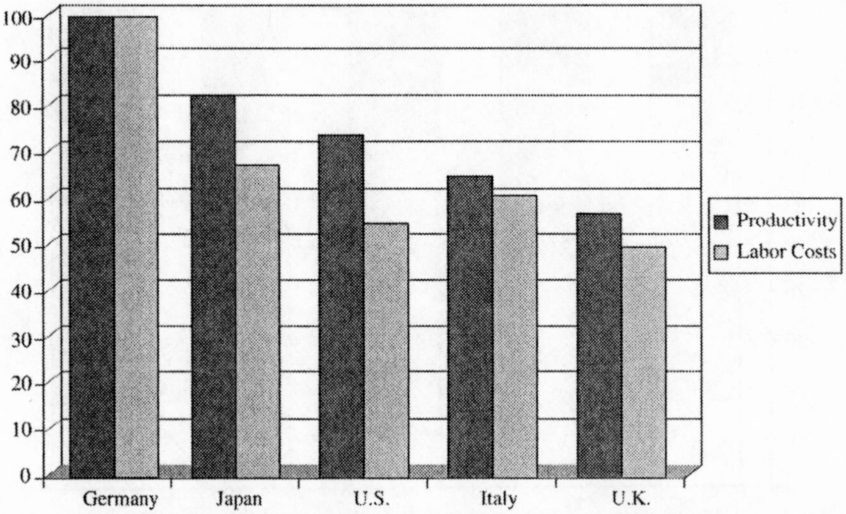

Compare *Institut der deutschen Wirtschaft Koln* (ed.): 1998—Zahlen zur wirtschaftlichen Entwicklung der Bundesrepublik Deutschland, Koln 1998.

fields, such as biotechnology, gene technology, microelectronics, microsystem technology, and telecommunication technology. On this point, German research facilities and organizations need to catch up.

Figure 14.5 shows the state of production technology of German plants compared to the other countries in the study. It indicates that 80 percent of the German organizations rank their technologies as better than their global competitors.

A further factor is the use of modern machinery. With 17 percent of the machinery less than 2 years old, German plants are in the lead (Figure 14.6). Unlike Japanese and U.S. plants, German plants have very few old machines in use. Only 8 percent of the German machinery is older than 20 years old. In addition, German plants have a balanced relationship between the age groups of their equipment.

The rate of process improvement is often accelerated by production employees. In German plants, there is a strong emphasis on providing incentives for improvement. In terms of the use of nonfinancial incentives for improvement suggestions, Germany is similar to Japan and the United

Figure 14.5
Production Equipment Relative to Their Industry

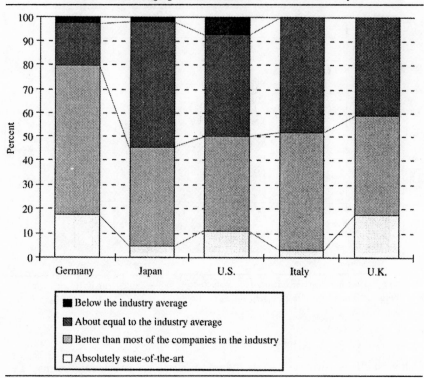

Source: HPM data.

States; however, German plants dominate the other countries in their use of financial rewards for improvement suggestions.

Demand Conditions

International competitiveness also depends on demand conditions. Segmentation and quality of domestic demand are important for early recognition and interpretation of customer needs and, therefore, for the creation of customer-relevant product features. The strength of domestic demand forces organizations to achieve continuous improvement and innovation.

A further competitive factor is the size of the national market as defined by the number of inhabitants and their economical potential. For foreign investors, the size of the German market is a genuine reason to

Figure 14.6
Age of Equipment in Plants

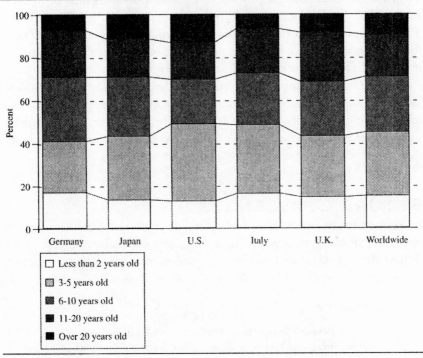

Source: HPM data.

invest in German operations. Even where there are relative cost disadvantages, this statement is true. Thus, in the past, organizations had little incentive to actively internationalize, due to the size of the German home market. Mobile and multinational native customers contribute to the internationalization and the stimulation of the external demand by exportation of domestic demand.

As shown by their export rates, German products enjoy a high international demand. The importance of exports to German organizations is shown by the fraction of the exports in the gross national product, which is predicted to increase to 35 percent in the year 2000. Japan, with exports a predicted 14 percent of its gross national product for the year 2000, and the United States, with 18 percent, illustrate the importance of exporting products and services for Germany.

Corporate Strategy, Structure, and Competition

To investigate their strategies, we asked respondents to assess their own products compared to the products of their strongest competitor (see Figure 14.7). They were asked about the sale price of the product, the percentage of sales for research and development, the percentage of sales for distribution and marketing, product quality, brand image, and product features.

Of the German plants, 56 percent assess their products as better-than-average or superior. Brand image and product quality were evaluated similarly. Only some of the plants describe superiority in price; however, there was a larger percentage of German plants that described price superiority than there was for plants in other participating countries.

FUTURE CHALLENGES

In addition to the factors described up to this point, the state can influence international competitiveness. The government can influence factor conditions through taxes and subsidies, capital market restrictions, spending

Figure 14.7
Product Competitive Performance Comparison
(Percentage of German Plants That Rate Their Plant as
Better-than-Average or as Superior to the Competition)

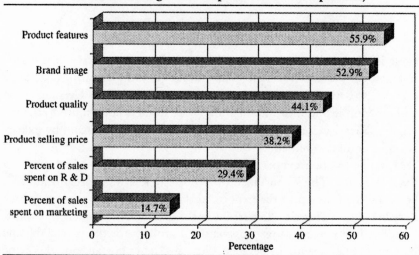

Source: HPM data.

on infrastructure, or training. Demand conditions are reinforced by security standards and by public purchasing and material management. Finally, barriers to new competition and the German Antitrust Law vary the situation and structure of a plant's competition.

Additionally, events which organizations have no control over determine the evolution of their international competitiveness. Such events can include technological ruptures, inventions, drastic changes in demand, or commodity prices. An example of such an event that has had a major impact on demand for many organizations was the opening of the eastern boundaries. The markets have enlarged themselves, on the one hand, on account of the new German states, and on the other hand, on account of the aperture of the East European countries.

A further event that may influence international competitiveness of German organizations is shifts in world financial markets and international exchange rates. With the start of the European Monetary Union, markets within Europe are becoming more transparent, and European organizations are finding themselves confronted with a changed competitive situation. Especially in financial markets, the monetary infrastructure of European competitors will serve the companies with a high liquidity grade and low transaction cost, a prerequisite for a healthy competitive environment throughout Europe. It will also strengthen Europe's competitive advantage in the triad with Japan and the United States and will prepare Europe and the European Union for strong global competition in the twenty-first century.

NOTE

1. We have used data from the "old" states of the former Federal Republic only, since on the average the competitive situation in the "new" states can still not be compared to World Class Manufacturing standards.

PART IV

CONCLUSIONS

CHAPTER 15

CONCLUSIONS AND THE WAY FORWARD

ROGER G. SCHROEDER and BARBARA B. FLYNN

This is a unique book dealing with manufacturing from a global perspective. It brings together an international team of experts from the United Kingdom, Germany, Japan, Italy, and the United States to address common issues related to high performance manufacturing (HPM). This has resulted in a global perspective about issues facing manufacturing firms today.

This book is based on the most comprehensive database about manufacturing plants in the world today. The database, spanning five countries and 164 plants, provides valid cross-country comparisons and has resulted in fresh insights concerning manufacturing practices and performance. We have drawn on this database and our plant visits to develop and support the findings in this book.

The scope of this study and of the book is very broad. It includes not only strategy in manufacturing plants but also a wide range of practices, including just-in-time (JIT), total quality management (TQM), human resources, technology, and information systems. As a result we take a broad view of practices that can improve manufacturing (see Figure 15.1).

CONTINGENCY APPROACH

Some of our findings are quite surprising. First, we have observed dramatically different paths being followed in different countries over the past 40 years. For example, beginning in the 1960s, Japanese plants started with "soft" innovative practices such as employee involvement, JIT, and TQM. In Japan this was followed by "hard" innovative practices, such as computer-aided design and manufacturing (CAD/CAM) and automation, in the

Figure 15.1
HPM Model

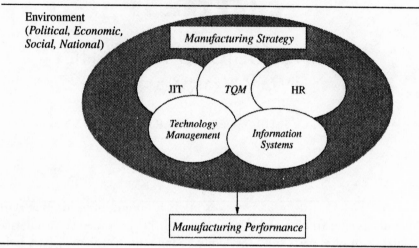

Environment
(*Political, Economic,
Social, National*)

Manufacturing Strategy

JIT   TQM   HR

*Technology
Management*   *Information
Systems*

Manufacturing Performance

1980s and 1990s. In Germany, however, the reverse situation occurred, starting with hard technology first and adding soft practices later.

We also discovered that some high-performing plants have chosen selectively from the set of all possible practices, whereas others have implemented nearly all practices that come along. Our analysis of performance indicates that both the "selector" strategy and the "comprehensive" strategy can work. Plants do not need to implement every new practice that comes along—they can be selective in their choices. On the other hand, they can take the comprehensive approach if it fits their particular situation and strategy.

Based on our research, we found that high-performing plants use different approaches, based on their size, strategy, country, industry, and other factors that describe their situation or context. We have, therefore, concluded that a contingency approach to adoption of practices is appropriate, rather than a "one-size-fits-all" approach. While we recommend that plants should choose practices that fit their strategy and situation, we also caution them to not be complacent and use this as an excuse to avoid change.

To show how this contingency approach works, consider one plant that we visited in the United States. This plant was a producer of relatively innovative electronic products. The plant had been growing in sales at 20 percent per year but was still a relatively small plant, compared to its competitors, with only 200 employees. The plant used a repetitive production process and produced most of its orders to stock. Quality of conformance and on-time delivery to customers was very good, but the plant faced cost problems that had been aggravated by foreign competitors, falling prices, and lack of aggressive management in the plant. As a result, margins were eroding in the business, causing pressure on the plant to reduce its costs.

What approach should the plant take to increase its performance? Plant management considered JIT, automation, process improvement, and further employee involvement (EI) as possible approaches to reduce costs. Employee involvement was somewhat questionable because the unionized workforce had been critical of management and of new approaches aimed at cutting jobs and "co-opting" the union. More automation might be effective, but upper management had indicated that the plant must improve its performance first before further investment would be made. Plant management referred to this as the "show me the money" approach from corporate.

The plant competed on service to its customers, in terms of delivery times and innovation of new products. Thus, in reducing costs, it did not want to jeopardize the current competitive strategy. Cost reduction was viewed as a necessary move to increase margins, rather than a move toward a new low-cost manufacturing strategy. With this in mind, the plant felt that implementation of JIT would be the appropriate approach to reducing costs and meeting its other strategic objectives. This was determined after benchmarking the plant against some its best competitors and similar plants in other industries. After JIT had been successfully implemented, continuous improvement of processes would be undertaken in order to prevent similar margin erosion in the future. The plant also expected new automation to be provided after JIT was successful and the financial results were demonstrated.

Note how the selection of practices in this case are guided by manufacturing strategy and sequenced over time. At the initial point in time, certain practices are ruled out, such as EI and automation, even though the plant may be behind in these areas. Ruling these out is consistent with selecting an appropriate path of improvement based on strategy, context, and the situation facing the plant. Had the plant been located in another industry or country, the choice of the best path might have been different.

The approach presented in this book is clearly different from others. We do not endorse the concept of universal best practice. Best practice, as we see it, must be subjected to fit before adoption. However, even if plants adopt different practices to fit their particular situation, they can end up at roughly the same place over time. This is particularly true of the comprehensive practice adopters in the same industry and situation. So when we observe that "everybody is doing it," that may be correct for some practices that indeed have become universal over time. Practices adopted and the order of implementation should be put to the contingency test.

LINKAGES ACROSS PRACTICES

The second major point in this book is the notion of linkages across practices. When comparing two plants with the same set of practices and the same situation, we find differences in performance due to the extent of linkage among practices. For example, one linkage that must be carefully managed is the linkage between strategy and the practices selected.

It is critical that plant improvement approaches be linked to the business strategy and manufacturing strategy. Strategies should be held constant over time, if possible. Nevertheless, plants in rapidly changing industries, where strategies and competitors are constantly changing, face a difficult problem. For example, a plant should not implement hard automation and make other investments that take years to pay off when the plant environment is rapidly changing. Rather, the plant should seek flexibility when facing a turbulent environment. In effect its strategy should be one of flexibility.

It is not enough to merely select the appropriate strategy; we found that communication of strategy and implementation is the key to high performance. A plant must also understand how its capabilities can lead to strategies that meet existing customer needs or that help to penetrate new markets.

In addition to strategy, the linkages across practices are critical. For example, we found a common infrastructure underlies JIT, TQM, total productive maintenance (TPM), and other practices. This common infrastructure often consists of human resources (HR) practices such as teamwork, cooperation, cross training, and incentive systems. Unless the appropriate infrastructure is in place, all other practices will provide less than their full potential. In addition, use of the appropriate infrastructure will allow pursuit of simultaneous dimensions of competitive performance.

We observed the situation of trapped linkages in Chapter 11. This occurs when communication and action processes are weak or missing. A plant can achieve a leveraged linkage by strong communications across the plant and implementation of new practices, directed by its strategy. Linkages are trapped when practices are not appropriately linked together through strong communications or actions taken.

Another example of linkage that we found in our studies was in the area of information systems. Information systems that are designed to support the practices and processes in place will lead to higher performance than packaged or canned approaches. Today, we see an alarming trend toward the use of standard packages to support information systems. For example, many plants have adopted standard enterprise resource planning (ERP) systems or packaged approaches to ERP. While this may be prudent in certain situations, managers must carefully assess how the standard package will support unique processes and practices that the plant is implementing. If a standard package does not fit, the linkage will be trapped or missing.

In the technology area, we also note the need for linkages. Product and process technologies must be linked, as we have argued in Chapter 6. Also, the manufacturing technology must be in step with the HR system and information systems used. If linkages are weak or missing, the technology investments are likely to be less effective and performance will be reduced.

Returning to the plant example that we discussed earlier, the important linkages now become clear. Starting with the decision to implement JIT to shorten cycle time and to reduce cost as the preferred approach, there are many other practice linkages that must be made. One of these is cross training, which, in turn, affects selection, training, and retention of employees. Another linkage is between setup-time reduction for JIT and product design, equipment selection, and so on. The point is clear: the effectiveness of JIT implementation will depend on how well it is linked with other practices and functions inside and outside manufacturing.

HPM: An Elusive Target

The third major message in this book is that HPM is an elusive target. Sometimes we find that plants can actually have high performance with relatively poor practices. In this case, the plant can rapidly regress in performance when conditions change, since a solid foundation of practice is not in place. This can be like the situation of complacency that we described in Chapter 12.

We also found situations where plants have implemented a high level of practice and still perform poorly. These plants are underachievers; they need to consider whether they have chosen the right practices for their situation and whether the practices are appropriately linked with strategy and each other. Implementation status is depicted by the "football" diagram of practice and performance in Figure 15.2.

In our analysis of plants we found a high level of variation in plant performance. Only a few plants in our study were able to achieve "world class" status in both performance and practices. It is, therefore, necessary that a plant carefully diagnose its situation and then set out on a deliberate path of continuous improvement. A plant must not rest on its laurels or be overly confident in its success. This is, of course, easier to say than to do, leading to an explanation of why HPM is so difficult to achieve and sustain in actual practice.

One of the problems that plants can encounter, and we have cautioned against, is "death by a thousand initiatives." In one company in the United Kingdom, for example, a new manager arriving on the scene found 197 improvement initiatives in place. Not surprisingly, people in the plant talked about initiative overload and initiative fatigue. The new manager was able to focus the plant on those initiatives relevant to the plant's strategy and to choose those initiatives that fit the plant's situation and reject those that did not.

Figure 15.2
The Five Starting Points for Improvement

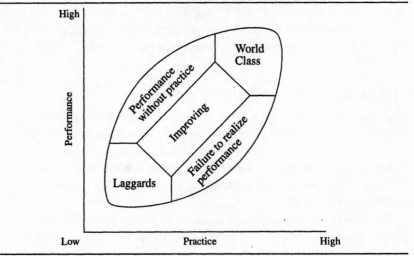

Another case we described in Chapter 12 was the Mercedes automobile plant where high performance was being achieved in the mid-1980s, with relatively poor practices. At that time, quality was being inspected into the product, and Mercedes was reported to be devoting as many labor hours to rework as the Japanese used, in total, to assemble their luxury cars. The product quality was good at Mercedes, but the practices were poor. This results in poor cost performance because high quality is being achieved only by maintaining a large "hidden plant" of employees and other resources to inspect and to rework poor-quality items.

Plants on the top may have particular trouble staying there for long. We detected an equalizing force in play. Plants that are behind are more aggressive than the top plants, in many cases. This can lead to role reversal where the laggards become the leaders. It may be harder to stay on top than to get there.

GLOBAL IMPLICATIONS OF HPM

Because this book is on global manufacturing, we would be remiss to end without summarizing some cross-country comparisons. U.S. manufacturing has caught up with Japanese manufacturing in many areas and surpasses it in some cases. This has been done by retooling U.S. industry and implementing "Japanese" approaches some of which were originally invented in the United States. Macroeconomic conditions have also helped the United States catch up. The combination of aggressive management and a favorable economy demonstrates the equalizing force in this case.

Many of the German plants that we studied are also aggressive world class competitors. Germany has demonstrated a flair for technology, equipment, and engineering. More recently, it has added social democracy and "soft" practices, such as cross training, TQM, and JIT, thereby demonstrating a uniquely German path toward HPM.

Even the United Kingdom has shown remarkable resilience in manufacturing. Once the cradle of manufacturing in the world, the U.K. plants have rebounded by implementing leading-edge practices when they fit their strategy. They have linked practices, and they seek to emphasize their particular strengths in the area of new product design.

Italian plants have managed to improve, despite economic obstacles and a small local market. This has led Italian manufacturing to be very export oriented and to be innovative in the practices adopted. Benetton, for example, in northern Italy is an innovator in supply chain management from suppliers through to customers. It continues to pursue its strategy of networking, while controlling the supply chain.

Finally, Japanese plants have shown their resilience to very tough economic conditions in the past 10 years. They have emphasized linkages and strong measures to control manufacturing costs. At the same time, Japanese plants have invested more in research and design (R&D) and innovation aimed at new product introductions. They have not given up their leadership in the world manufacturing arena; rather, plants in other countries have caught up with the Japanese "manufacturing miracle" of the 1970s and 1980s.

THE PATH TO HPM

In this age of fast information exchange and benchmarking, it is difficult to gain an advantage for long by implementing innovative new practices. What seems to be critical is the linkages established among practices and the set of practices selected that fits the plants particular situation.

What will the future bring? Forecasting is a dangerous occupation, as most economists can attest. Nevertheless, new techniques and practices are sure to emerge. No one can tell what these new techniques and practices will be, only that they are bound to arrive, probably at an increasing pace. After all, who would have predicted 20 years ago that we would have lean manufacturing, supply chain management, ERP, e-business, or Six Sigma approaches today?

This book provides a guide for selection of practices that fit your company. Rather than touting yet another new practice, it provides a road map for choice and implementation. Being aware of new practices and fitting them to the situation, while linking the new and the old, is the path to high performance manufacturing.

Appendix

This appendix contains articles published by members of the HPM group using data collected from the common database. Many of these articles are research papers and provide support for conclusions drawn in the various chapters.

Ahmad, S. "The Relationship between JIT Managerial Practice and JIT Infrastructure: Implementation for Plant Performance," Ph.D. Thesis, University of Minnesota, 1998.

Anderson, J., M. Rungtusanatham, and R. Schroeder. "A Theory of Quality Management Underlying the Deming Management Method." *Academy of Management Review* 19, no. 3 (July 1994).

Anderson, J., M. Rungtusanatham, S. Devaraj, and R. Schroeder. "A Path Analytic Model of a Theory of Quality Management Underlying the Deming Management Method: Preliminary Empirical Findings." *Decision Sciences* 26, no. 5 (1995).

Banker, R., S. Devaraj, K. Sinha, and R. Schroeder. "Performance Impact of the Elimination of Direct Labor Variance Reporting: A Field Study." *Journal of Accounting Research* (2000).

Banker, R., G. Potter, and R. Schroeder. "Manufacturing Performance Reporting for Continuous Quality Improvement." *Management International Review* 33, no. 1 (1993): 60–85.

Banker, R., G. Potter, and R. Schroeder. "An Empirical Analysis of Manufacturing Overhead Cost Drivers." *Journal of Accounting and Economics* 18 (1994).

Bates, K., R. Filippini, C. Forza, and A. Vinelli. "Personnel and Customer Satisfaction: A Cross Country Study." In *Managing Service Operations: Lessons from the Service and the Manufacturing Sectors,* edited by J. Ribera and J. Prats, 507–513. IESE, Canon Editorial, 1997.

Bates, K., S. Amundson, W. Morris, and R. Schroeder. "The Crucial Interrelationship between Manufacturing Strategy and Organizational Culture." *Management Science* 41, no. 10 (1995).

Bates, K. "Manufacturing Strategy Implementation Relationships with Performance, Plant Culture and Plant Structure," Ph.D. Thesis, University of Minnesota, 1995.

Bertorelle, P., C. Forza, and A. Vinelli. "Trade off: Storia o Realtà per L'impresa di Fine Millennio?" *Finanza, Marketing e Produzione* 16, no. 3 (1998): 7–45.

292   APPENDIX

Bertossi, G., C. Forza, and A. Vinelli. "Il Ruolo delle Risorse Umane nella Strategia di Customer Satisfaction." *Sviluppo e Organizzazione,* no. 157 (Settembre/Ottobre 1996).

Cua, K.O. "A Theory of Integrated Manufacturing Practices: Relating TQM, JIT, and TPM." Ph.D. Thesis, University of Minnesota, 2000.

De Toni, A., M. Muffatto, G. Nassimbeni, and A. Vinelli. "Evoluzione dei Rapporti fra Grandi Imprese Operanti in Mercati Internazionali e Micro Imprese Subfornitrici Locali," [Evolution of relationships between big companies operating in international markets and small local subcontractors] *Sinergie,* no. 36–37 (Gennaio–Agosto 1995).

De Toni, A., M. Muffatto, G. Nassimbeni, and A. Vinelli. "Performance and Organization of Very Small Firms in the Supply Chain." In *Proceedings of the 30th International Matador Conference,* edited by A.K. Kochhar. Manchester, England: MacMillan Press, 31, March–2, April 1993.

De Toni, A., M. Muffatto, G. Nassimbeni, and A. Vinelli. "Supply Polices of Large and Medium Firms: Empirical Finding and Comparisons." In *Proceedings of the International Symposium on Logistics,* edited by K.S. Pawar, 6–7. Nottingham, England: The University of Nottingham Press, July 1993.

De Toni, A., and G. Nassimbeni. "Buyer-Supplier Operational Practices, Sourcing Policies and Plant Performances: Results of an Empirical Research." *International Journal of Production Research* 37, no. 9 (1999): 597–619.

De Toni, A., and G. Nassimbeni. "Just-in-Time Purchasing: An Empirical Study of Operational Practices, Supplier Development and Performance." *OMEGA,* forthcoming.

De Toni, A., R. Panizzolo, G. Nassimbeni, and A. Vinelli. "Service Concepts in World Class Manufacturing." In *Productivity and Quality Management Frontiers,* edited by D.J. Sumanth, J.A. Edosomwan, R. Poupart, and D.S. Sink. Norcross, Georgia: Industrial Engineering and Management Press, 1993.

De Toni, A., C. Forza, and R. Filippini. "Just-in-Time and Time Performance: Reference Model and Empirical Results." In *Operations Strategy and Performance,* edited by K.W. Platts, M.J. Gregory, and A.D. Neely. Cambridge: Cambridge University Press, 1994.

De Toni, A., R. Filippini, and C. Forza. "Manufacturing Strategy in Global Markets: An Operations Management Model." *International Journal of Operations and Production Management* 12, no. 4 (1992): 7–18.

Ebine, A., and M. Morita. "Quality Competitiveness and Communication Systems." *Journal of the Japan Society for Management Information* 8, no. 2 (September 1999): 13–39 (in Japanese).

Filippini, F., C. Forza, and A. Vinelli. "Human Resource Development in Customer Satisfaction Strategies." Decision Sciences Institute Proceedings, San Diego (22–25, November 1997): 1519–1521.

Filippini, R. "Operations Management Research: Some Reflections on Evolution, Models and Empirical Studies in O.M." *International Journal of Operations and Production Management* 17, no. 7: 655–670.

Filippini, R., P. Romano, and A. Valeri. "Strategie di Produzione e Sviluppo di Tassonomie. Una Ricerca Empirica Internazionale." *Finanza Marketing e Produzione*, Anno XVI, no. 4: 131–186.

Filippini, R., C. Forza, and A. Vinelli. "Improvement Initiative Paths in Operations." *Integrated Manufacturing Systems* 7, no. 2 (1996): 67–76.

Filippini, R., C. Forza, and A. Vinelli, "Sequences of Operational Improvements: Some Empirical Evidence." *International Journal of Operations and Production Management* 18, no. 2 (1998): 195–207.

Filippini, R., C. Forza, and A. Vinelli. "Il Trade Off tra Prestazioni in una Prospettiva di World Class Manufacturing." [Performance Trade-Off in World Class Manufacturing]. In *Leve Strategiche nei Mercati Integrati*, [Strategic levers in integrated markets] pp. 175–196, edited by A. La Bella, M. Raffa, and G. Zollo. Milano: Etas Libri, 1995.

Filippini, R., C. Forza, and A. Vinelli. "Trade Off and Compatibility between Performance: Definitions and Empirical Evidence." *International Journal of Production Research* 36, no. 12 (1998): 3379–3406.

Flynn, J., B. Flynn, R. Filippini, C. Forza, A. Vinelli, and R. Schroeder. "Trade-offs versus Synergies in Manufacturing Performance Dimensions." *Midwest Decision Science Annual Meeting.* Indianapolis, Indiana (24–26, April 1997).

Flynn, J., B. Flynn, R. Schroeder, R. Filippini, C. Forza, and A. Vinelli. "The Relationship between Manufacturing Practices and Performance Trade Offs and Compatibilities." *Decision Science Institute Conference,* Orlando, Florida (24–26, November 1996).

Flynn, J., and K. Bates. "Innovation History and Competitive Advantage: A Resource-Based View Analysis of Manufacturing Technology Innovation." *Best Papers Proccedings, Academy of Management* (1995).

Flynn, J., and K. Bates. "Firm Innovation History and Competitive Advantages: A Resource-Based View Analysis of Manufacturing Process and Technology Innovation." *Proccedings of the Annual Meeting of the Decision Sciences Institute* (1994).

Flynn, B.S., and E.J. Flynn. "The Relationship of Strategic Management to World Class Manufacturing." *Proccedings of the Annual Meeting of the Decision Sciences Institute* (1990): 1302–1304.

Flynn, B., S. Sakakibara, and R. Schroeder. "The Relationship between Quality Management Practices and Performance: Synthesis of Findings from the World Class Manufacturing Project." *Advances in the Management of Organizational Quality,* edited by Donald Fedor and Soumen Ghosh, vol 1. JAI Press, 1996.

Flynn, B.B., and E.J. Flynn. "Information Processing Alternatives for Coping with Manufacturing Environment Complexity." *Decision Sciences* 30, no. 4 (1999).

Flynn, B.B., and E.J. Flynn. "The Relationship between Simplification Alternatives and Manufacturing Performance. *Proceedings, European Operations Management Association,* Venice, Italy (1999).

Flynn, B.B. "The Relationship between Quality Management Practices, Infrastructure and Fast Product Innovation." *Benchmarking for Quality Management and Technology* 1, no. 1 (1994): 48–64.

Flynn, B.B., K.A. Bates, R. Schroeder, S. Sakakibara, and E.J. Flynn. "World Class Manufacturing in America." *Proceedings, 1989 Decision Sciences Institute Annual Meeting* (1989): 880–882.

Flynn, B.B., E.J. Flynn, R. Filippini, C. Forza, A. Vinelli, and R. Schroeder. "Configurations of Compatible Dimensions of Competitive Performance: An International Analysis." *Proceedings, Decision Sciences Institute International Meeting,* Athens, Greece (1999).

Flynn, B.B., S. Sakakibara, and R.G. Schroeder. "Relationship between JIT and TQM: Practices and Performance." *Academy of Management Journal* 38, no. 5 (1995): 1325–1360.

Flynn, B.B., S. Sakakibara, and R. Schroeder. "The Relationship between JIT, Quality and Plant Performance." *Proceedings, 1992 Decision Sciences Institute Annual Meeting* (1992): 1409–1411.

Flynn, B.B., S. Sakakibara, and R.G. Schroeder. "The Interrelationship between JIT and TQM: Practices and Performance." *Academy of Management Journal* 38, no. 5 (1995).

Flynn, B.B., S. Sakakibara, K.A. Bates, and E.J. Flynn. "Symposium: World Class Manufacturing and Plant Performance" (symposium). *Proceedings, 1991 Decision Sciences Institute Annual Meeting* (1991).

Flynn, B.B., R.G. Schroeder, and E.J. Flynn. "World Class Manufacturing Practices: An Empirical Investigation of the Hayes and Wheelwright Framework." *Journal of Operations Management* 17, no. 3 (1999): 249–269.

Flynn, B.B., R.G. Schroeder, and S. Sakakibara. "A Proposed Quality Management Theory and Associated Measurement Instrument." *Journal of Operations Management* 11 (1994): 339–366.

Flynn, B.B., R.G. Schroeder, and S. Sakakibara. "Determinants of Quality Performance in High and Low Quality Plants." *Quality Management Journal* 2, no. 2 (Winter 1995): 8–25.

Flynn, B.B., R.G. Schroeder, and S. Sakakibara. "Reliability and Validity Analysis of a Proposed Quality Management Measurement Instrument." *Best Papers Proceedings, Academy of Management Annual Meeting,* 1991.

Flynn, B.B., R.G. Schroeder, and S. Sakakibara. "The Impact of Quality Management Practices on Performance and Competitive Advantage." *Decision Sciences* 26, no. 5 (1995): 659–692.

Flynn, B.B., R.G. Schroeder, and S. Sakakibara. "The Relationship between Quality Management Practices and Performance: A Path Analytic Approach." *Proceedings, 1993 Decision Sciences Institute Annual Meeting,* 1993.

Flynn, B.B., R.G. Schroeder, S. Sakakibara, E.J. Flynn, and K.A. Bates. "The World Class Manufacturing Project: A Retrospective View of Research Issues and Methodology." *International Journal of Operations and Production Management* 17, no. 7 (1997): 671–685.

Flynn, E.J., and B.B. Flynn. "Achieving Simultaneous Competitive Advantages through Continuous Improvement: World Class Manufacturing as Competitive Strategy." *Journal of Managerial Issues* 8, no. 3 (1995): 360–379.

Forza, C., and F. Salvador. "Assessing Some Distinctive Dimensions of Performance Feedback Information in High Performing Plants." *International Journal of Operations and Production Management* 20, no. 3 (2000): 359–385.

Forza, C., and F. Salvador. "Information Flows in High Performance Manufacturing." *International Journal of Production Economics* (2000), forthcoming.

Forza, C., and A. Vinelli. "On the Contribution of Survey Research to the Development of Operations Management Theories." In *Operations Management: Future Issues and Competitive Responses,* edited by P. Coughlan, T. Dromgoole, and J. Peppard, 183–188. School of Business Studies, Dublin, 1998.

Forza, C., and F. Di Nuzzo. "Meta-Analysis Applied to Operations Management: Summarising the Results of Empirical Research." *International Journal of Production Research* 36, no. 3 (1998): 837–861.

Forza, C., and F. Salvador. "I Sistemi Informativi nelle Operations delle Aziende ad Elevate Prestazioni: uno Studio Empirico." *Finanza, Marketing e Produzione* 17, no. 4 (Dicembre 1999): 139–181.

Forza, C. "Achieving Superior Operating Performance from Integrated Pipeline Management: An Empirical Study." *International Journal of Physical Distribution and Logistic Management* 26, no. 9 (1996): 36–63.

Forza, C. "Quality Information Systems and Quality Management: A Reference Model and Associated Measures for Empirical Research." *Industrial Management and Data Systems* 95, no. 2 (1995): 6–14.

Forza, C. "The Impact of Information Systems on Quality Performance: An Empirical Study." *International Journal of Operations and Production Management* 15, no. 6 (1995): 69–83.

Forza, C. "Work Organization in Lean Production and Traditional Plants: What Are the Differences?" *International Journal of Operations and Production Management* 16, no. 2 (1996): 43–63.

Forza, C., and R. Filippini. "The Role of Information and Communication Systems in World Class Manufacturing." In *International Operations: Crossing Borders in Manufacturing and Service,* edited by R.H. Hollier, R.J. Boaden, and S.J. New, 175–180. Amsterdam: Elsevier Science Publisher, 1992.

Forza, C., and R. Filippini. "TQM Impact on Quality Conformance and Customer Satisfaction." *International Journal of Production Economics* 55, no. 1 (1998): 1–20.

Forza, C. *I Sistemi Informativi nella Produzione di Classe Mondiale,* [Information Systems in World Class Manufacturing], Tesi di dottorato [PhD Thesis], Istituto di Ingegneria Gestionale, Università di Padova, a.a. 1991–1992.

Hollingworth, D.G. "An Examination of Relationships Among Plant Performance Outcomes." Ph.D. Thesis, University of Minnesota, 1998.

Junttila, M.A. "Toward a Theory of Manufacturing Strategy." Ph.D. Thesis, University of Minnesota, 2000.

Maier, F.H. "Competitiveness in Manufacturing as Influenced by Technology—Some Insights from the Research Project: World Class Manufacturing." In *Systems Approach to Learning and Education into the 21st Century,* Yaman Barlas, Vedat G. Diker, and Seçkin Polat, Editors, Vol. 2, 667–670. Istanbul, 1997.

Maier, F.H. "Competitiveness of German Manufacturing Industry—An International Comparison." In *Proceedings, Decision Science Institute—Vol. 3 POM—Manufacturing.* E. Powell Robinson, David L. Olson, and Benito E. Flores, Editors, 1171–1173, San Diego, 1997.

Maier, F.H. "Feedback Structures Driving Success and Failure of Preventive Maintenance Programs." In *Proceedings of the 7th International Annual EurOMA Conference,* Ghent, Belgium, June 4–7, 2000.

Maier, F.H., P.M. Milling, and J. Hasenpusch. "Implementation and Outcomes of Total Productive Maintenance." In *Operations Management—Future Issues and Competitive Responses,* Paul Coughlan, Tony Dromgoole, and Joseph Peppard, Editors, 304–309. Dublin, 1998.

Maier, F.H. "Technology: A Crucial Success Factor in Manufacturing?—Some Insights from the Research Project: World Class Manufacturing." In CD-ROM Proceedings of the International System Dynamics Conference 1998, Quebec City, Canada, 1998.

Marble, R.P., and F.H. Maier. "Inter-Organizational Information Sharing in Operations Networks: An Empirical Analysis of Its Influence on Manufacturing Performance." In Managing Operations Networks—European Operations Management Association Conference, Emilio Bartezzaghi, Roberto Fillipini, Gianluca Spina, and Andrea Vinelli, Editors, 19–26. Venice, Italy, 1999.

Matsui, Y. "Evaluating the Role of Manufacturing Department in Technological Development Activities: An Empirical Research for Machinery, Electrical and Electronics, and Automobile Plants in Japan." *Yokohama Business Review* 17, no. 4 (1997): 45–67. (in Japanese)

Matsui, Y. "Formulating Operations Strategy in Japanese Manufacturing Firms: An Empirical Research for Machinery, Electrical and Electronics, and Automobile Plants." *Yokohama Business Review* 19, no. 3 (1998): 16–46. (in Japanese)

Matsui, Y. "Total Quality Management in Japanese Manufacturing Firms: An Empirical Research for Machinery, Electrical and Electronics, and Automobile Plants." *Yokohama Business Review* 18, no. 4 (1998): 27–55. (in Japanese)

Matsui, Y. "Human Resource Management in Japanese Manufacturing Firms: An Empirical Research for Machinery, Electrical and Electronics, and Automobile Plants." *Yokohama Business Review* 20, no. 3 (1999): 45–75. (in Japanese)

Matsui, Y., and O. Sato. "An International Comparison Study on the Benefit of Production Information Systems." *Proceedings of Japan Society for Management Information* (November 1998): 185–188. (in Japanese)

Matsui, Y. "On the Utilization of Production Information System in Manufacturing Firms: An Empirical Research for Machinery, Electronics, and Automobile Plants in Japan." *Yokohama Business Review* 18, no. 2 (1997): 21–48. (in Japanese)

Matsui, Y. "On the Implementation of Just-in-Time Production System in the Japanese Manufacturing Firms." *Yokohama Business Review* 16, no. 4 (1996): 39–62. (in Japanese)

McKone, K., R. Schroeder, and K. Cua. "The Impact of Total Productive Maintenance Practices on Manufacturing Performance." *Journal of Operations Management* (2000).

McKone, K., R. Schroeder, and K. Cua. "TPM: A Contextual View." *Journal of Operations Management* 17, no. 2 (January 1999).

Milling, P.M., F.H. Maier, and D. Mansury, "Impact of Manufacturing Strategy on Plant Performance—Insights from the International Research Project: World Class Manufacturing." In Managing Operations Networks—European Operations Management Association Conference, Emilio Bartezzaghi, Roberto Fillipini, Gianluca Spina, and Andrea Vinelli, Editors, 573–580. Venice, Italy, 1999.

Milling, P., U. Schwellbach, Jörn-Henrik Thun, M. Morita, and S. Sakakibara. "Production Cycle Time as a Source of Unique Strategic Competitiveness—An Empirical Analysis Based on the World Class Manufacturing Project." In Proceedings of the 1st World Conference on Production and Operations Management, Sevilla, Spain, August 27–September 1, 2000.

Milling, P., U. Schwellbach, and Jörn-Henrik Thun. "Time as a Success Factor for Operations Management—An Empirical Analysis Based on the World

Class Manufacturing Project." In Proceedings of the 7th International Annual EurOMA Conference, Ghent, Belgium, June 4–7, 2000.

Milling, P., U. Schwellbach, Jörn-Henrik Thun, S. Sakakibara, and M. Morita. "Shortening Cycle Times by Developing an Environment for Fast Organizational Learning and Decision Making—An International Comparison Based on the WCM Project." In Proceedings of the 1st World Conference on Production and Operations Management, Sevilla, Spain. August 27–September 1, 2000.

Morita, M. "Effects of Information Technology and Organizational Communication." *Organizational Science* 29, no.1 (1995): 4–17. (in Japanese)

Morita, M. "Analysis of Management Structures: Systems Thinking and Strategy." In *Economic Analysis for Decision Making*, edited by M. Takahashi, H. Itami, and T. Sugiyama, 39–60. Yuhikaku, 1995. (in Japanese)

Morita, M. *Essence of Business Leaders.* Nikkei BP, 1997. (in Japanese)

Morita, M., and A. Ebine. "Human Resources Sustaining Excellent Company." *Human Resource Development* (Japan Management Association) 9, no. 12 (1997): 4–19. (in Japanese)

Morita, M., and A. Ebine. "Structuring of Practices for Speed Competence." In *Strategy-Driven Manufacturing: A Key for the New Millennium*, edited by H. Hiroshi, 177–182. Proceedings of the International Symposium on Manufacturing Strategy, 1998.

Morita, M., and E.J. Flynn. "The Linking among Management Systems, Practices and Behavior in Successful Manufacturing Strategy." *International Journal of Operations and Management* 43, no. 9 (1997): 967–993.

Morita, M., and S. Sakakibara. "Linkage as a Key for Excellence of Management, Part 1." *Management 21* (Japan Management Association) 4, no. 8 (1994): 48–52. (in Japanese)

Morita, M., and S. Sakakibara. "Linkage as a Key for Excellence of Management, Part 2." *Management 21* (Japan Management Association) 4, no. 9 (1994): 44–48. (in Japanese)

Morita, M., and E.J. Flynn. "Manufacturing Property as a Substratum for Effective Manufacturing Strategy." In *Manufacturing Strategy: Operations Strategy in a Global Context*, C.A. Voss, ed., 459–464. 1996 European Operations Management Association.

Morita, M., N. Tanaka, H. Mori, and Y. Takahashi. "Communication Network Systems for Competitiveness: The Japanese World Class Manufacturing Case." In *Proceedings of the 1995 System Dynamics Conference, Vol. 2 (Plenary Program)*, edited by K. Saeed and Toshiro Shimada, 150–169. The System Dynamics Society, 1995.

Morita, M., R. Filippini, and E.J. Flynn. "The Capability of Linking Practices to Create Strategic Leverage." In *Managing Operations Networks*, edited by E. Bartezzaghi et al. 877–884. European Operations Management Association, 1999.

Nakamura, M., S. Sakakibara, and R. Schroeder. "Adoption of Just-in-Time Manufacturing Methods at U.S. and Japanese Owned Plants: Some Empirical Evidence." *IEEE Transactions on Engineering Management* 45, no. 3 (August 1998).

Nakamura, M., S. Sakakibara, and R. Schroeder. "Japanese Manufacturing Methods at U.S. Manufacturing Plants: Empirical Evidence." *Canadian Journal of Economics* 29 (1996).

Nassimbeni, G. "Factors Underlying Operational JIT Purchasing Practices." *International Journal of Production Economics* 42, no. 3 (1996): 275–288.

Rungtusanatham, M., C. Forza, R. Filippini, and J.C. Anderson. "A Replication Study of a Theory of Quality Management Underlying the Deming Management Method: Insights from an Italian Context." *Journal of Operations Management* 17, no. 1 (1998): 77–95.

Sakakibara, S., B.B. Flynn, and R.G. Schroeder. "A Framework and Measurement Instrument for Just-in-Time Manufacturing." *Production and Operations Management* 2, no. 3 (1993): 177–194.

Sakakibara, S., B.B. Flynn, R.G. Schroeder, and W.T. Morris. "The Impact of Just-in-Time Manufacturing and Its Infrastructure on Manufacturing Performance." *Management Science* 43, no. 9 (1997): 1246–1257.

Sakakibara, S., B.B. Flynn, and R.G. Schroeder. "A Just-in-Time Manufacturing Framework and Measurement Instrument." *Production and Operations Management* 2, no. 3 (1993).

Sakakibara, S., R.G. Schroeder, and B.B. Flynn. "Japanese Manufacturing Management: A Three-Cycle Model." *Proceedings, European Operations Management Association Annual Conference,* London (June 3–4, 1996).

Sakakibara, S., B.B. Flynn, R.G. Schroeder, and W.T. Morris. "The Impact of Just-in-Time Manufacturing and Its Infrastructure on Manufacturing Performance." *Management Science* 43, no. 9 (1997).

Sakakibara, S., M. Rungtusanatham, E.J. Flynn, and R. Luechtefeld. "Just in Time, Decentralization, and Manufacturing Performance: An Empirical Investigation of Their Relationships." *Proceedings of the Annual Meeting of the Decision Sciences Instutute* (1993): 131–136.

Salvador, F., C. Forza, M. Rungtusanatham, and T.Y. Choy. "Interactions Across the Supply Chain: Toward Theoretical and Empirical Development," edited by E. Bartezzaghi, R. Filippini, L. Spina, and A. Vinelli. *Managing Operations Networks,* SGE Editoriali (1999): 27–34.

Sato, O., and Y. Matsui. "International Comparison of Information Technology Usage in Factories." *Proceedings of the 36th National Meeting of Japan Society for the Study of Office Automation,* pp. 113–116, October 1998. (in Japanese)

Sato, O., Y. Matsui, and K. Chou. "Implementation Problems of CIM." *Office Automation* 14, no. 2 (1993): 89–92. (in Japanese)

Schroeder, R., B.B. Flynn, E.J. Flynn, and D. Hollingworth. "Manufacturing Performance Tradeoffs: An Empirical Investigation." *Proceedings, European Operations Management Association Annual Conference,* London, June 3–4, 1996.

Schroeder, R.G., B.B. Flynn, and E.J. Flynn. "An Empirical Investigation of the Hayes and Wheelwright Framework." *Proceedings, Decision Sciences Institute Annual Meeting,* 1997.

Schroeder, R.G., B.B. Flynn, S. Sakakibara, E.J. Flynn, and K.A. Bates. "Empirical Analysis of World Class Manufacturing" (symposium). *Proceedings, 1990 Decision Sciences Institute Annual Meeting,* 1990.

Schroeder, R.G., S. Sakakibara, E.J. Flynn, and B.B. Flynn. "Japanese Plants in America: How Good Are They?" *Business Horizons* (July–August 1992): 66–72.

Tuerk, K. Informations Systeme der Produktion und ihre Unterstützung durch Gruppenarbeit zur Steigerung der Wettbewerbsfähigkeit—Eine Empirische Untersuchung im Rahmen des Projektes World Class Manufacturing (Written in German, English title, Information Systems in Production and their Support Through Teamwork—Empirical Results of the Project World Class Manufacturing), Ph.D. Dissertation Mannheim University 1998, Publisher: Duncker und Humblodt, Berlin, 1999.

Weissmann, S., Erfolgsbeitrag von Praktiken eines Umfassenden Qualitätsmanagements für Industriebetriebe—Eine Empirische Untersuchung im Rahmen der Internationalen Studie "World Class Manufacturing" (Written in German, English title: Success Factors of Total Quality Management Practices for Industrial Enterprises—Empirical Results of the International Study World Class Manufacturing), Ph.D. Dissertation Mannheim University, 2000.

INDEX

Printed in the United Kingdom
by Lightning Source UK Ltd.
103434UKS00001B/98